2008 SUBMARINE ALMANAC

Also by Deep Domain Publishing
2007 Submarine Almanac

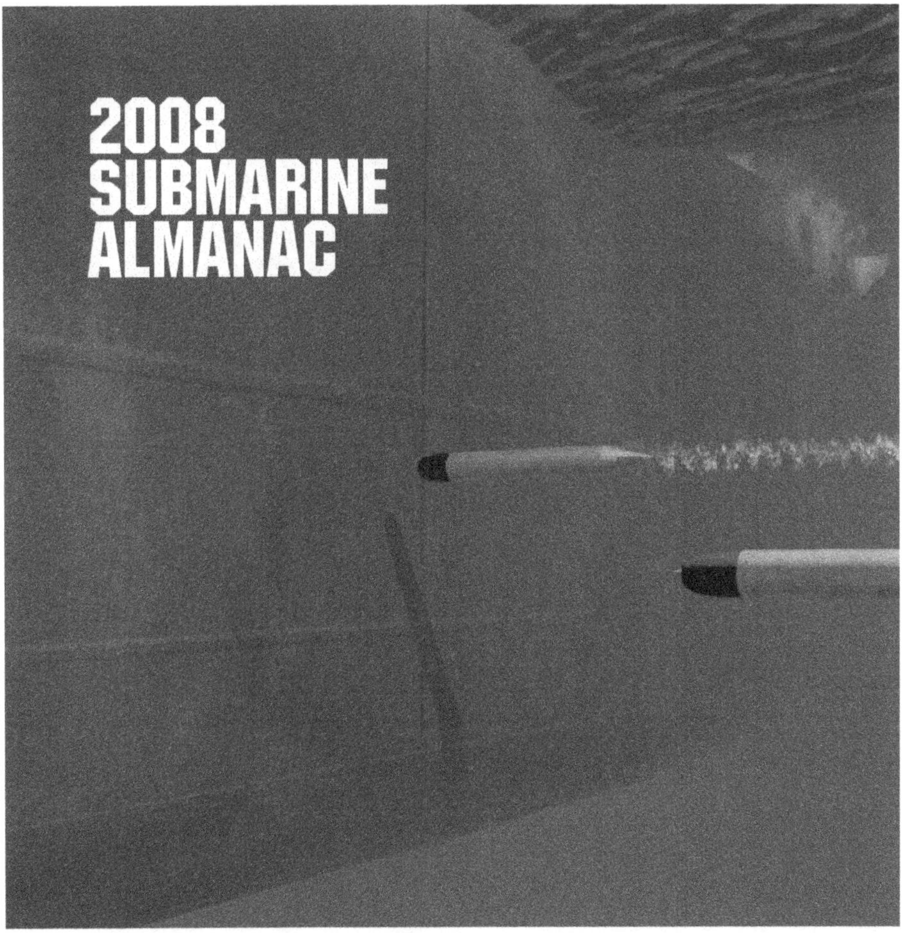

2008 SUBMARINE ALMANAC

NAVAL STORIES, SUBMARINE HISTORY, AND GAME DEVELOPMENT FROM EXPERTS AND ENTHUSIASTS

Neal Stevens, Editor

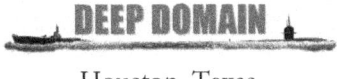

Houston, Texas

2008 Submarine Almanac
Copyright © 2008 by Neal Stevens
Subsim ~ www.subsim.com

All rights reserved. No part of this book shall be reproduced, stored in a retrieval system, or transmitted by any means, electronic, mechanical, photocopying, recording, or otherwise, without written permission from the publisher. Brief excerpts are permitted as the subject of reviews of this book.

Cover art and arrangement by Spencer Burnham
Back cover *USS Texas* by Marko Barthel

First Edition
ISBN 978-0-6151-8426-5
Printed in the United States of America

For Valerie, Vickie, & Natasha

Published by the Deep Domain
ww.deepdomain.net
Houston, Texas USA
2007

Foreword

Submarines always have, and always will, do it for me. The allure of a vessel able to disappear below the tossing waves of a hostile sea and sail invisibly toward a mission has captured my imagination since I could walk.

When my classmates were playing with model airplanes, I could be found in the bathtub with a submarine model, either a German U-boat or a sleek 1950s *Seawolf*-hulled sub, my junior diver mask on so I could see the sub approaching periscope depth. I surrounded myself in books about submarines, and tried to find out about nuclear subs, but little was available. Other than the famous journey of the *Nautilus* to the polar icecap and some mention of nuclear ballistic missile subs, there was little in my childhood libraries about modern submarines.

I was fortunate to win an appointment to Annapolis. The second summer there, I had a week-long cruise on a fleet ballistic missile nuclear sub carrying Poseidon missiles, the follow-on to the Polaris ICBMs and predecessor to the Tridents. I was assigned a coffin-like bunk by missile 11. At a briefing in the crew's mess for us 'riders', the captain explained that during midshipmen operations, we had a fallback mission in case of nuclear disaster. In the event of a devastating nuclear strike against America and NATO, the ship would quietly and slowly slip to deeper water, evade detection, and three months after the nuclear world war was lost, launch its sixteen multiple reentry vehicle warheads at predetermined targets inside the Soviet Union.

The ship was truly a doomsday device. I raised my hand and asked how the crew would feel about launching more nukes on a planet already laid waste by them. Would the crew cooperate and launch, knowing that the result would just be more death? The captain's jaw tightened and there may have been a quiver in his voice as he replied, "Son, most of us have wives and young children at home. If Russia goes nuclear and wipes us out, you can goddamn bet that every single member of this crew will live for the moment that we can exact our revenge. We will follow our orders, even if the men who wrote those orders ceased to exist months ago. That's our mission and our purpose."

Later, much later, as the communications officer of a fast attack nuclear submarine, I was in charge of the top-secret-material safe, which meant constantly updating the books within, the publications or simply the 'pubs'. There was no way to do this without reading snippets of top-secret war plans as I replaced an old page with a new one, signed off that I'd done so, then shredded the old page and signed off that the destruction was complete. During long, dull Saturday nights in port, as duty officer, I would open the safe and read the pubs, and I learned the other half of the doomsday plan that is in large part responsible for saving civilization.

You see, our FBM ballistic missile submarines were quieter than a hole in the ocean. They couldn't be detected from afar, and they could disappear if an intruding, trailing enemy submarine followed them from up close. We knew that secret personally by conducting security exercises, 'SecExes', or 'sucking eggs' as we derisively called them. These exercises pitted an American attack submarine against an American ballistic missile submarine to see if the attack boat could detect and trail the missile boat.

You have to know that we in the attack submarine force felt like we were the fighters while the FBM boats (the 'boomers') were the bombers. We were agile and swift and secret. We were 'forward deployed' to spy on enemy coastline activity, to monitor communications, to tap into underwater phone cables and intercept secret messages, to deploy commandos offshore and bring them back, and surreptitiously follow enemy nuclear subs as they went about their sinister missions.

One secret mission I experienced proved the daring of SSN attack boats. We penetrated deep inside Libya's 'Line of Death' in the Gulf of Sidra, where a huge Soviet fleet lay at anchor in that Cold War battleground, the Mediterranean. For us the Med was an ideal place to station our fleet because it was in easy weapons range of European Russia, while the Russians deployed there because it was in range of the heart of NATO. At the anchorage was our target, a *Victor* class Russian nuclear attack submarine, a quiet, bad-ass, loaded-to-the-gills with ship-attack cruise missiles and aircraft carrier killing torpedoes with the mission of trailing the *USS America* task force with missile tubes open, ready to take out our carrier.

Our mission was to find the *Victor* in the anchorage and wait for him to start his engines and put out to sea, and when he submerged, we'd follow him with *our* torpedo tube doors open and our weapons locked onto him.

And if he made one false move, we'd take him out. Even in the Cold War, shooting a nuclear submarine was something that might be acceptable because so many things could go wrong submerged that we could have explained it away as a Russian accident. That was one reason we had the Glomar Explorer, not just to haul up the occasional sunken Russian sub hull for its intelligence harvest, but to cover up the evidence in the cases when we shot first and asked questions later. If there were no wreckage, there could be no accusations that we had sunk a Soviet submarine.

The trouble was, the Russians were waiting for us to try and trail their submarine. In the anchorage was a vast Russian nuclear-powered battleship – they called her a cruiser – named *Kirov*, loaded with every conceivable weapon a warship could carry, including submarine-killing depth charges and torpedoes. *Kirov* was armed with a huge sonar capable of reaching out ten miles at sea with active pinging, and even further with passive listening, to detect intruding submarines. As we crept up on the Soviet flotilla, slowly moving to avoid coming up on the Russian sonar's 'Doppler alarms' and scraping the bottom so that our submarine would just look like the ocean floor on the Russian sonar screens, the *Kirov* was shrieking her sonar deep into the gulf, 'illuminating' the entire bay with extremely loud and powerful sonar beams.

The sound of the *Kirov's* sonar was terrifying. It blasted into the hull, and not only could it be heard with the naked ear, it was so loud that we had to yell over it. Even if the Soviets couldn't detect us, their 'Death Ray' sonar beams were the ultimate in undersea psychological warfare. For days we withstood the Death Ray sonar as we sneaked up on the anchorage, and when we got there, we went to the one place of safety – *directly underneath the hull of the* Kirov! Five feet beneath that battle cruiser, the sonar was pinging too close for its computers to interpret the return ping. Had we extended our periscope, it would have banged onto the bottom of *Kirov's* hull. We waited, occasionally thrusting sideways out from under the *Kirov* to raise the periscope, under darkness, and take photos of the assembled Russian warships and determine the bearing to the anchored and shut-down Russian *Victor* submarine.

One ship in the flotilla was a gray-painted cruise ship. I remember commenting on it, saying, "What the hell is a new cruise ship doing in the Russian order of battle?" None of our intel had such a hull. The captain smiled at me, saying, "That's their 'comfort ship'. It's a huge floating hotel, casino, sports club and bordello. The crews take a launch over to the cruise ship, get their share of R-and-R, then go back to minding the store on their combat vessels."

"Captain," I whined, "can't *we* go to the comfort ship?" After over forty days submerged without a break, in the company of 130 sweating guys, the thought of a cold beer and a hot Russian prostitute was captivating.

After what seemed forever, the *Kirov* turned off his Death Ray sonar. And an hour later, the *Victor* started his reactor and steam plants, rigged for dive, weighed anchor and sailed north to catch up with the *USS America*. And for the next forty days and forty nights, we shadowed him, and he had no idea we were there, a football field behind him, with two torpedoes locked on at all times.

So with some justification, we attack boat sailors thought ourselves somehow cooler than the calm gentlemen of the boomer force who had the unglamorous task of hiding. "Hide with Pride," the boomer sailors would grin as they rolled dice to see where random chance would tell them to steer the ship, another roll to see which operation area they would orbit in, their location unknown even to their command structure, as they ran silently away from oil tankers and fishing boats to keep their locations and their missiles' locations secret.

But the SecExes mentioned above, an attack boat versus ballistic missile boat exercise, proved who the real professionals were. We SSN attack boat sailors would start the exercise a few hundred yards behind, 'in trail', of an SSBN boomer ship, with the job of trying to see if we could shadow them out to sea. If we could trail the boomer for an hour, we'd win. In SecEx and after SecEx, we'd pursue the boomer with all our might. Most of the time we'd lose the bastards as they faded silently away into the deep.

Sometimes we'd triumphantly claim we'd been able to trail the boomer for the requisite hour, only to find out that ten minutes into the SecEx, the boomer had launched a device from a torpedo tube, a device that moved like him and sounded like him, while he went quiet and waited for us to go by, pursuing a decoy. The lesson – even our super-quiet attack boats were no match to try and trail our own ballistic missile boats.

And with our 'acoustic advantage' over the Russians (we were quieter with more sensitive sonars, allowing us to sneak up on them or detect them before they could detect us), that meant that if *we* couldn't hear our ballistic missile submarines, neither could the Russians. And *that* meant that our ballistic missile submarines were truly invisible and undetectable. They were the perfect deterrent force.

Why do I say they won the Cold War? Wasn't it just the weight of the Soviet Empire that made them collapse on their own? Or Reagan's insistence on bluffing them with Star Wars space weapon systems?

No! The answer was buried in that top-secret reading material, the War Plan. The document that stated exactly what we were to do if 'the balloon went up', if the Russians launched a first strike and caught us with our pants down.

We needed no orders. In the best case we'd get a one-line radio message telling us to execute the plan. If not, we'd execute it anyway. The War Plan was a 'break glass in case of emergency' publication. It functioned as sealed orders should anything go wrong. In chapters within the plan, depending on what assignment our ship had that week, we would go to that section and it would tell us where to go and what our mission was. For example, most of our assignments had us following a 'submarine safe lane' to the northern Arctic coast of Russia, where we were to sink their Northern Fleet as they sortied from the Kola Peninsula. Our prime targets were the Russian ballistic missile submarines, then the Russian fast attack subs, and then their aircraft carriers and cruisers.

All that seems obvious and unsurprising. But the interesting chapters concerned the ballistic missile subs. Their orders, depending on what their assignment was that week, was either to launch on predetermined targets or wait, sometimes weeks, sometimes months, and strike long after the Russians thought they'd won. In some cases, our ballistic missile subs were to become agents of psychological warfare by launching one missile a week at random targets.

What was different about our submarines' deterrent force from the Air Force's silo-based missiles and bomber-based weapons, and the Army's short-range missiles and projectile nuclear weapons, was PALs, preventive action interlocks. The Air Force and Army guys needed their nukes 'unlocked' by Washington, or else they were inert or even capable of self-destructing. The reason – their nukes could be stolen, whereas ours were locked up in missile tubes inside a steel hull inside a guarded Navy base or deep underwater at sea. The Navy's weapons had no PALs.

This meant that the captain and crew of a nuclear submarine were on the same war response level as the president, at least in the case of America being the victim of a surprise nuclear first attack. The War Plan said it in black and white. If a boomer crew lost their 24/7/365 receive-only radio signal from Washington, they were instructed to come to periscope depth and listen to commercial frequencies. If the captain became convinced that America had suffered a nuclear Pearl Harbor, he had the authority *on his own* to launch a retaliatory strike at Russia.

It was a real life Dr. Strangelove-esque doomsday device. But as in Dr. Strangelove, what good is a doomsday device if the other side didn't know about it?

That was the mission of our friends in the Defense Intelligence Agency, to leak parts of the top-secret War Plan to the Russians. The Soviets didn't believe what they read in the *Washington Post* or the *New York Times*. Only what they gleaned from paid spies deep within America's and NATO's defense structure. That is what famous double agent spies like Aldrich Ames and John Walker did for us.

While many believe that the spy scandals of the 1980s were a case of the incompetence of America's intelligence agencies when their spies were found selling secrets to the Russians, I believe that a big part of this was a deliberate coordinated effort to let the Russians know that attacking America or NATO was a complete losing proposition.

Once the Soviets knew the score, there was no sense in continuing the arms race and the Cold War. Pragmatism and reason prevailed, even in the heart of communist Russia. The Berlin Wall came down and the Soviet Union was disassembled, and in an attempt to save face, the Russians acted as if it were all their idea.

With the Cold War safely over and the Russian menace tamed, civilization could continue. All thanks to the American and NATO submarine forces. But what now? Do we still need the expense and trouble of a nuclear submarine fleet in the post-Cold War era in which our battles are with rag-tag stateless terrorists?

It is a new era of warfare. In this century, we may live to see the day that a container ship carries a nuclear, chemical or biological weapon into the port of Baltimore, or at least tries to. The intelligence gained by means such as surreptitious offshore spying submarines could do what agents on the ground and satellites in orbit can't, and with these present and future threats of city-killer cargo ships, submarines are perfect for putting them on the bottom long before they approach our shores.

While that is one possible scenario, we've seen that in every chapter of modern warfare, the submarine has been a key element in victory. Odds are that new challenges will show further usefulness of submarines that exceeds even the imagination of a fiction writer. But until the last submarine is chopped into razor blades, submarines will always be the most amazing part of the defense establishment, and will always be my passion.

Read on, and you'll be sure to agree with me.

Michael DiMercurio
Author of *Vertical Dive*
silentfastdeep@aol.com

Introduction

Why do we love submarines?

The slender stalk of a periscope, alien, spying on the world above the waves, suggesting a greater malevolence lurking close below. The crowded conning tower, packed with desperate men, silent as if their lives depended on it, the hostile pinging of the enemy sonar calling out to them. The skipper, one eye pressed to the scope, arm leisurely looped around a handle, the only man able to actually see what is going on around them, urgently calling out bearing, range, and angles to the attack party. The officer feeding the information into the Torpedo Data Computer, a marvel of pre-transistor technology, blending electro-mechanical magic and science into a firing solution. The torpedo, patient, lethal, awaiting the opening of the outer door, surging without mercy toward the target.

In popular culture, no machine of war comes close to matching the mystery and allure of the submarine. For thousands of years men have breached the barrier between surf and shore with craft that plowed through the waves, but never below them. As ships progressed from rafts and canoes to triremes and galleons, to steamboats and dreadnoughts, all nautical invention and history were still literally scratching the surface. The submarine is the sole and true master of the sea.

The love affair with submarines began before they actually existed. David Bushnell's *Turtle*, Horace Hunley's self-named *H.L. Hunley*, and Bourgeois' *Plongeur* were not submarines but concepts brought to life, brief flirtations into the third dimension of sailing. Jules Verne's *Nautilus* gave the world a glimpse of the staggering power of a true submarine but men would have to wait for ingenuity to catch up with imagination.

When technology evolved to yield the propeller, diesel engine, high-capacity storage battery, and riveted steel hull, the age-old dream of a submersible ship truly began to take shape. After WWII, nuclear propulsion finally allowed the submarine reality to match the vision. The modern submarine is able to hide from the eyes of the world, to appear and strike without any possible warning, and to escape undetected within the folds of the ocean where no surface ship impeded by that nautically immature characteristic of positive buoyancy can follow.

Sure, as enthusiasts of naval ships, we cherish the great 18th-century sailing ships and all they represented: man's full courtship of the sea, the brave voyages of exploration, the leaps of technology in navigation and armament, and the storied sea battles they fought. We appreciate the fury of the massive battleships and the combined air/naval might of the carrier. Still, no vessel has generated a following like the submarine—books, movies, computer games; all overwhelmingly favor the sub.

In 1998, I was invited for a day cruise aboard the *USS Houston* SSN-713, a 688-class nuclear attack sub. Like any ship, *Houston* left the dock and sailed out to sea, with lookouts posted on the bridge to stay clear of maritime traffic. As I stood on the bridge, I could see fishing boats, a US warship, and some merchant vessels all going about their business. Nothing in this scene was remarkable. Then the captain ordered *Houston* to dive and we descended into the control room, closed the hatches, and slipped beneath the waves, leaving the surface inhabitants behind.

During our submerged run we simulated torpedo attacks, visited the sonar shack, and had a wonderful Tex-Mex dinner. Then came the exercise joyfully referred to as "angles and dangles". For the next forty minutes *Houston* and her crew showed what she could do, tearing through the ocean depths at greater than 25 knots, sailing on the z-axis, from periscope depth to greater than 800 feet. As I held tightly to the lines along the bulkhead, I had a mental picture of this 7000-ton, bullet-black submarine, twisting and turning, surging through the water like a true sea beast, while above her bobbed a fishing boat, its small engine languidly puttering along, the crew working the nets in the calm sea breeze and sun, unaware of the war machine racing through the abyss below them.

That's why we love submarines!

Neal Stevens
Houston, Texas
December 1, 2007

CONTENTS

Almost Successful - Mariano Sciaroni with J. Matthew Gillis	19
Cinematic Sub Stuff - Bob 'Dex' Armstrong	29
Why Submarines Are Better Than Women - Mike Hemming	41
Der Drache ist Todt - Grant Swinbourne	43
Submarines from Containment to Preemption - Capt. Zeb Alford	74
Looking Back at Fast Attack - Jim Frantz	82
The Krusanov Ultimatum - Andrew Glenn	88
Morning Lookout – Kevin Moffat	104
Puppies of the Pacific - Chris Weisensel	105
Do You Believe in Miracles, Jake? - Alan Bradbury	118
Growl, Tiger - Ron Gorence	133
The Center of the World - Mike Hemming	138
Submarine Dictionary	149
Submariner Speech from WWII to Present - Tammy L. Goss	155

A Sub and Crew Worthy of the Name Texas - Neal Stevens ·········· 164

Silent Hunter II Memoirs - Shawn Storc ························ 170

Tales from the Torpedo Room - Don Meadows ················· 179

The Diary of a U-boat Commander - Sir Stephen King-Hall ········· 190

USS Casimir Pulaski: Story of a Cold War Warrior - Don Murphy ····· 305

Subsim Roll Call ·· 313

Contributors ··· 319

2008 SUBMARINE Almanac

"Just one torpedo can ruin your whole day."

Almost Successful
ARA *San Luis* War Patrol

Mariano Sciaroni

with J. Matthew Gillis

Argentine Submarine Force Coat of Arms

ARA *San Luis* is most famous for serving in the Falklands War. After HMS *Conqueror* had sunk the cruiser ARA *General Belgrano*, the Argentine fleet retired to port for the duration of the war, with the exception of the *San Luis*. She was the only Argentine naval presence facing the British fleet.

The IKL 209 / 1200 Submarines

The Armada de la República Argentina (ARA) submarine *San Luis* (S-32) was one of the two new Type 209 submarines in service in 1982 with the Argentine Navy. These vessels were also designated *Salta* class after *San Luis'* sister submarine.

Both were ordered in 1968 and, starting in the latter half of 1969, were built in sections by Howaldtswerke Deutsche Werft (HDW) AG, Kiel, West

Germany. The individual sections were then shipped to Argentina for assembly at Tandanor, Buenos Aires, and the two submarines were finally commissioned in 1974.

IKL Industries of Lübeck was responsible for the design of the Type 209 class, which was based on units of smaller displacement (around 450 tons submerged) corresponding to the older Type 205 and 206 classes constructed for the Bundesmarine (then the West Germany Navy). It is noteworthy that the 209 series began with four submarines requested by Greece, followed by the Argentine order. The ARA ships displaced 1,200 tons submerged, slightly larger than their Greek counterparts.

The acquisition, assembly, and operation of the new IKL 209/1200s submarines represented a qualitative jump for the ARA, which, until the date of their acquisition, had only operated older Guppy submarines entering the final phases of their life spans. After a few years the Guppies would be replaced by the *Santa Cruz* (TR 1700) class of submarines that were being constructed in Germany at the time.

Type 209 ships were different from previous ones operated by the submarine branch of the ARA, with advanced electronics, improved surface search radar, more capable active/passive sonar suite, modern fire control and combat information center, and an advanced electronic surveillance system (ESM).

A Rude Awakening

April 2nd, 1982, the Falkland Islands conflict between the United Kingdom and Argentina ignited. The drums of war surprised the British. No less surprised was ARA *San Luis* and all of its crew, including her commander, docked at a pier of Mar del Plata Naval Base (BNMP), home base of the small ARA submarine force.

The surprise came because the Argentine Naval High Command chose to keep the operation initiated and finalized that day secret. They did not inform the commanders of the diverse units not directly involved. The operation initially designated 'Blue' and later (mid-approach to target) 'Rosary' was the assault on the Malvinas/Falklands Islands.

However, ARA *San Luis* did not receive the order to get ready for a combat patrol until twenty-four hours after the assault of the Malvinas/Falkland. The order issued by the Submarine Forces Command (COFUERSUB) numbered as 2/82 COFUERSUB instructed the *San Luis* to "disrupt the operations of the British Expeditionary Force in the

Malvinas–Georgias focal area in order to contribute to maintaining and consolidating the conquest of the Malvinas."

The TOAS (South Atlantic Theater of Operations) commander, an 'old school' submarine officer, Vice Admiral Lombardo, had reasoned that a conventional diesel/electric submarine could not safely go forth to seek an encounter with the enemy, but instead "needs to be in a patrol area and to wait for it there."

In fact, whether a submarine is in an offensive or defensive role is determined by its 'indiscretion time'; the time she is exposed above the water's surface. A submarine on an offensive mission must travel faster and therefore must snorkel, surface transit, and recharge batteries more frequently than one defending an area.

In response to this reasoning, Lombardo's staff selected three large patrol areas for the *San Luis*: around the Georgias Islands, near Ascension Island, and in the proximity of the Malvinas/Falkland. The final choice was for the Malvinas/Falkland when a study considered the area more suitable than the Georgias, having too few targets, and Ascension Island, being too far away.

The State of Affairs

At that moment, the *San Luis*' crew began to ready the ship in order to make her combat capable in the shortest possible time. The initial state of the ship was not entirely satisfactory, and she needed to enter drydock (something that would have to be made in the Puerto Belgrano Naval Base, the main base of the Argentine Navy because Mar de Plata lacked the necessary facilities).

ARA *San Luis*' hull, propeller and refrigeration pipes had build-ups of small parasitic crustaceans, which affected her performance and increased her noise level. Because there was not enough time to travel to Puerto Belgrano, the cleaning was done in Mar del Plata by divers who worked twenty-four hours a day for nearly a week.

In addition to these problems, the unit presented other deficiencies not entirely corrected in time for departure, even with the heavy work done by the crew and civilian contractors. One diesel engine was out of service because the engine block was broken, and the three others had problems that limited their output power. Seawater routinely entered from the snorkel, and she also had breakdowns in the bilge pumps.

The training level of the crew was affected by the personnel rotation policy of the Navy—many crewmembers were new to the ship—but was 'acceptable' in the opinion of the unit's skipper, Frigate-Captain (a rank nonexistent in the U.S. Navy below captain and above commander) Fernando Azcueta.

However, various key posts in the submarine, such as those in the fire control systems, were occupied by junior NCOs. The disorganization brought complications later in the combat patrol (most veteran submariners in service in the Argentine Navy were, at the time, in Germany, where they were supervising the construction of the new TR-1700s). Furthermore, while *San Luis*' command crew had experience in submarines, neither the skipper nor the second-in-command had any experience with the Type 209s.

Problems on the Way

On April 11th, in late-afternoon hours, not a minute later than the scheduled departure time, ARA *San Luis* left from Mar del Plata for the Malvinas/Falkland area of operations. She was loaded to the top with water and food, and armed with ten SST-4 anti-surface wire-guided torpedoes and fourteen Mk. 37 antisubmarine acoustic homing weapons.

Still unknown to many people today, it was not the first time the submarine visited these cold South Atlantic waters. In order to test the capacities of the new IKL 209/1200s submarines, a patrol in the neighborhood of the Malvinas/Falkland Islands was ordered by Navy High Command in 1975. The patrol was completed satisfactorily, and it included a night on the seabed of a bay in the proximity of the island's capital.

On April 17th after a transit without further problems, during which the skipper organized time to further train the crew and to order smaller pending repairs, the radio operator received a ciphered message. The orders were for the ARA *San Luis* to head to a waiting zone designated 'Enriqueta', located to the east of San Jorge's Gulf, near Argentina's mainland.

Two days later the Signaal M8/24 Fire Control Computer was broken beyond the repair capabilities of the junior NCOs in charge of its operation. The fire control system is the brain of an attack submarine. Fed by the diverse sensors searching the outside ocean, it allows the submarine to resolve a fire control solution, in order to shoot straight-running or homing torpedoes. The M8/24 underwater fire control system can simultaneously track and prepare fire control solutions for three targets, and control three torpedoes aimed at those targets. The system could be used not only to calculate torpedo lead angles but also to process all sensor data to give target positions and vectors. It could simultaneously display sonar, radar and periscope range and heading data.

It was a devastating loss. The loss of the fire control computer left the submarine without her automatic fire capability and limited her to just one torpedo which could be guided in manual (also called 'emergency') mode by the crew.

Nevertheless, the Submarine Force Commander acknowledged the problem and decided against retiring ARA *San Luis* from the combat area. The decision was made after evaluating the necessity of having at least one submarine on patrol even with the tactical limitations imposed by *San Luis'* inactive fire control system. ARA *Salta* had problems that made her inoperative and ARA *Santa Fe* (the old Guppy still in service) was on her way to the Georgias, where she was later chased and disabled by Royal Navy antisubmarine helicopters.

By April 26th negotiations for the fate of the Islands were nearly finished, and COFUERSUB decided to send ARA *San Luis* to the 'Maria' patrol area, located to the north of the islands, where she arrived by the 28th.

On the following day, with the Argentine military and political situation deteriorating, ARA *San Luis* received the order to destroy any enemy target located within the total exclusion zone around the Malvinas/Falkland: *Weapons free.*

The First Shots

In the early hours of May 1st, the *San Luis'* sonar apparatus, an Atlas Elektronik CSU 3, detected in passive mode the distant hint of a contact. It was quickly classified as either a Type 21 or Type 22-class frigate. Helicopter noise was also detected along the contact bearing.

The detected frigate was likely one of a group of two ships, frigates HMS *Brilliant* and HMS *Yarmouth*, and the helicopter noise was probably produced by Sea King helicopters from 826 squadron. The enemy was deployed near the area designated for the Argentine submarine, with the objective of pursuing her on authority of British commander Admiral Sandy Woodward, acting on reports offered by the British intelligence. Those reports had been submitted after an interception of an Argentine signal: the message directed from Mar del Plata to ARA *San Luis'* skipper ordering the sub to the 'Maria' patrol area.

After a silent approach maneuver, and at a distance about 10,000 yards to the target, Frigate-Captain Azcueta ordered the manual launching of an SST-4 anti-surface torpedo. Two minutes after the torpedo launched (Type 209 submarines have a 'swim out' torpedo tube system), the Weapons Control Coordinator slammed his fist on the console and in frustrated voice informed Azcueta that no signal was being received from the torpedo, indicating it had cut the guidance wire and lost all further contact with the submarine. No explosion was heard, and the final destination of the torpedo remains unknown.

It was not clear if the British forces detected the Argentine submarine after the unsuccessful attack, although the captain of the *Brilliant* specifically confirmed it later. However, no counterattack was launched specifically on the Argentine sub.

Anticipating an attack that never arrived, ARA *San Luis* settled on the bottom, remaining there for several hours. While the sub was resting on the seabed, the British hunter/killer party was chasing sonar contacts everywhere but in the *San Luis*' immediate area. HMS *Brilliant* sank at least two whales that day using torpedoes, while HMS *Yarmouth* attacked other contacts, firing more than thirty depth mortars. Also, the Sea King helicopters dropped several aerial torpedoes and depth charges on false targets, probably additional biological contacts. The expenditure of 203 British antisubmarine weapons produced no hits.

The fault in the torpedo weapons system affected the crew's moral, and in Azcueta's words, "to know the torpedoes did not explode made me feel a great impotence… it's one hard experience, I felt a big frustration."

The SST-4 Torpedo Failures

Much has been written about the ARA *San Luis*' torpedo failures. The torpedo carried by the *Salta* class subs was the SST-4 'Seal', of German manufacture: 533 mm in diameter, weighing 1414 kg (including a 260 kg warhead), and could be fired at a maximum distance of 40,000 yards at surface targets.

The problems with the SST-4 had been known in the Argentine Navy since 1981, with the initial versions of the weapon (those carried by ARA *San Luis*) known to suffer serious faults and lacking in effectiveness, although the causes of the problems were not known. Obviously, these reports were challenged by the manufacturer, AEG. Immediately after the war AEG reported that the lack of results in the torpedo attacks happened because the submarine "fired her torpedoes at too great a distance (in excess of 8000m), and using only passive bearings to estimate the range."

The same report, without any direct knowledge in the matter, also stated that a "synchro misalignment had caused incorrect bearing information to be transmitted from the periscope to the fire-control console." Furthermore, an "overzealous leading petty officer...had incorrectly reconnected leads used to power-up torpedoes in their tubes before launch."

But the many theories of AEG, pointing to ARA *San Luis'* crew and dismissing the torpedo faults, were proved wrong. Serious post-war studies made by AEG in the Baltic Sea showed that the original SSTs were unable to maintain depth while under wire control. They later repaired, without much fuss, most of the exported torpedoes, including the ones in Argentine service. It is important to note the problems on the SST-4s could not be solved during the war, although to date, they remain totally operational in the Argentine Navy.

Kill the Whales!

By May 4th, ARA *San Luis* was informed of the successful attack on the Type 42 destroyer HMS *Sheffield* (by the Super Etendard – Exocet weapons system), and she was ordered to the attack area in order to make a battle damage assessment and to hit targets of opportunity.

The order was revoked just a short time later, and thus she did not approach the area where HMS *Sheffield* was attacked. According to this information, it is clear the multiple sightings of periscopes by HMS *Glasgow* and HMS *Yarmouth* in the zone where Sheffield was burning were not more than optical illusions, whales and other forms of marine life or, perhaps, some Soviet submarine, as some sources suggested.

Adding to the mayhem, according to the British Task Force commander's words, "*Yarmouth* thought they heard a torpedo in the water and broke off to try and find the submarine that had fired it. Then it happened again. And again. All together they thought they detected nine torpedoes that afternoon. Sometime later we deduced that the propeller noises they kept hearing on the sonar had come from the outboard motor which was buzzing around *Sheffield* helping to fight the blaze."

Four days later, on May 8th, the Argentine submarine was trying to identify a faint contact when the sonar operators heard to a strange noise in the stern sector which was quickly evaluated as a possible torpedo. Trying to avoid the weapon's lock, the submarine was forced to make evasive maneuvers and to drop countermeasures.

Upon regaining the contact, and after sonar ruled out the possibility of marine life producing it, the skipper concluded it was a submarine on approach. At 21:42 he ordered the firing of a Mk. 37 antisubmarine torpedo, and a detonation was heard sixteen minutes later on the target's bearing, possibly after the torpedo striking the sea floor. No secondary explosions were detected and the sonar operators lost the initial contact after the detonation. Post-war analysis concluded the target of ARA *San Luis'* torpedo was possibly a whale.

Certainly, a big problem in any ASW operation is ascertaining the true identity of a given underwater contact. It generally requires extended tracking on multiple sensors to determine whether or not a sonar contact is a submarine or some other underwater phenomenon. The period of classification can be extraordinarily tense for the ship or submarine concerned during peacetime exercises, but in war the tension heightens and frequently elicits a higher false-contact rate. Both navies involved in the war can testify to that.

Below the Sea

At this point in the combat patrol, the moral of *San Luis'* crew, even with all the misfortunes, was good and remained the same for the duration of the patrol. One must remember, living inside the Argentine sub (as in any other diesel/electric boat in a combat patrol) was not an easy task. Lack of cooking (to avoid depleting the batteries), lack of fresh water, a bitter cold atmosphere (the temperature within the submarine was a constant 2°C, like the sea outside), and many others circumstances were enemies of the submariner's comfort and routine.

Yet the men were totally confident of the abilities of skipper, Azcueta, and Azcueta was very confident in his crew. The camaraderie of the submarine warriors helped during the hard moments below the South Atlantic.

Firing in Anger

By May 10th, in late-afternoon hours, ARA *San Luis* was on guard in the northern entrance to Estrecho de San Carlos (Falkland Sound) when she detected a surface contact entering the sound. She positioned herself to wait for it, sure it would return.

About 00:30 on May 11th, the sonar operators in the sub began to track not one but two contacts on their consoles. According to its noise signature, the submarine classified it as two ships, either destroyers or frigates, and began her attack maneuver (no Argentine ship was active in the patrol area so any contact was assumed hostile).

These warships were the Type 21 frigates HMS *Arrow* and HMS *Alacrity*, returning to the main British nucleus at 22 knots and with streamed torpedo decoys after having penetrated the Falkland Sound. At their high speed, their sonar equipment was unable to detect the slow approach of ARA *San Luis*, who, cruising underwater, approached to within 5,600 yards of the following target -- the HMS *Alacrity* -- and shot an SST-4 torpedo.

Just as had happened on May 1st, a few minutes after firing the weapon the sub received a signal indicating the torpedo had cut the wire, missing both targets (though some sources claim the lost torpedo struck the torpedo decoy being towed by HMS *Arrow*).

The attack went unnoticed by the British warships. In fact, Captain Chris Craig, HMS *Alacrity's* skipper, learned of the frustrated attack only a year later upon reading a battle report about ARA *San Luis*.

The crew manning ARA *San Luis* expected a heavy antisubmarine chase. In the planesman's words, "We think…okay, we are done…now they (the British) are going to start their counterattack." But nothing happened. Because of the target's speed and assuming incorrectly that both ships were at the time alerted, Captain Azcueta broke off the attack.

A few hours later, a disheartened ARA *San Luis* skipper broke radio silence and transmitted to his superiors the following message: "At Ensenada del Norte Position, I have attacked two DDs/FFs, first torpedo fired in emergency mode cut the wire, no impact. Launching annulled on second target, I consider weapon systems non-reliable, own position known by the enemy."

After receiving the message, COFUERSUB ordered the unit to port. Finally, on May 19th, after thirty-nine days of intense combat patrol, ARA *San Luis* arrived at Puerto Belgrano. Although the crew and harbor personnel hurried the repairs in order to make the ship battle-ready again, less than a month later the Argentine forces surrendered in the Malvinas/Falkland Islands, ending the campaign before the repairs were finalized.

Even though the *San Luis* would not get a second chance, her crew had proven themselves in combat, clearly rising to the challenge.

⊕   ⊕   ⊕

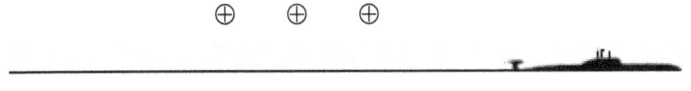

Epilogue

During all the hostilities, the British task force expended more than 200 ASW weapons against *San Luis* (or any contact they thought was the Argentine submarine) without scoring a single hit. Other than the destruction by a British aerial homing torpedo of a countermeasure decoy being towed by the British carrier HMS *Hermes*, the British ASW efforts were futile. Among these expenditures were more than thirty advanced (and very expensive) Mk. 46 air-launched torpedoes.

Although the war patrol of ARA *San Luis* did not produce losses on the British Task Force, it was surely not in vain, having tied up many resources of the British fleet in its hunt, resources which could have been used for other goals, such as naval bombing or radar picket duty. Great Britain committed five nuclear submarines, various destroyers and frigates, and at least four antisubmarine helicopters airborne at any time in the search of the solitary, stealthy ARA *San Luis*.

The elusive and determined Argentine submarine kept the British Task Force on full alert throughout all of the conflict, and because of this, limited its capacity of action. In fact, the submarine threat dictated much of the conduct of British naval operations during the campaign. Kept in this perspective, Frigate-Captain Azcueta and his hardworking crew totally fulfilled the given order to "disrupt the operations of the British Expeditionary Force in the Malvinas–Georgias focal area, in order to contribute, maintain and consolidate the conquest of the Malvinas."

For that reason, it is possible to conclude that ARA *San Luis'* war patrol was *almost* successful.

Cinematic Sub Stuff

Bob 'Dex' Armstrong

The German film *Das Boot* (The Boat) is the only film that closely approximates the boatservice I knew. Hollywood films never showed stores crammed everywhere, dirty laundry, skin books tucked above ventilation lines, gear adrift, piles of 'one-way' trash...and folks in raggedy-ass dungarees.

What they did show were officers wearing dress uniform hats in the barrel (conning tower). To get their eye up flush with the rubber eyepiece on the periscope, the officer's hat had to be turned around backward with the visor to the rear. Anyone ever see that? I never did. Anybody ever see a dirty cup in a submarine movie? How bout a full butt kit? A lookout wearing a straw hat? A mess cook in an apron that looked like it was salvaged from a leper colony dumpster?

How come the boats commanded by Cary Grant, John Wayne, Ronald Reagan, Clark Gable, Burt Lancaster, and Tyrone Power all operated using convent language? How come all the raghats went to fancy nightclubs full of knockout twenty-year-old blondes who drank Manhattans and danced to Glen Miller tunes in designer dresses? Where were all the hairy-lipped honeys that hung around the zoo cage bars we frequented?

How come all the boats returned freshly painted, and the happy non-rated guys all went bopping across the brow in fresh starched whites to be met by twenty wholesome bobby-soxers named Linda Lee and Peggy Sue?

Where were those admirals in dress canvas standing on the pier waiting to congratulate the old man when we came in?

How come the piers are all spotless and taxicabs are lined up to take all the squeaky-clean Arthur Murray qualified liberty hounds to the

Copacabana? How come no returning bluejacket ever meets some homely chick with six runny nose kids...a head full of curlers, worn out sandals and hands her a thirty-pound sack of filthy dungarees?

How come you never see some jerk hauling ass to get radio traffic and guard mail? Where do they hide the tenders with the rust stains? Speaking of rust, how come when highly paid Hollywood guys turn up on a pier, oxidation stops?

How come none of the returning drunks ever look like Ray Stone and Doc Beeghly? No missing clothing, blood, lipstick or leg chains.

How 'bout the nicknames on those Hollywood fleetboats? 'Rusty', 'Big Mike', 'Billy', or 'Smiling Eddy'? You never see anyone called 'Butt Face', 'Fat Ass', 'Fungus Foot' or 'Garbage Gut'. We had a kid nicknamed *The Chinese Whore*. We spent half an evening at a boat reunion trying to remember the kid's real name.

Who loads torpedoes and stores on those cinematic wonders? The Good Fairy? Shoemaker's elves? The entire tender crew out of the goodness of their brown-bagger hearts? The National Conclave of the Little Sisters of the Poor?

There is always a scene where Mr. Admiral Warmheart has Captain Cleanliving in his office.

The admiral speaks...

"Jack, I've got to give you a rough one this time."

"Bingo Lizard Straits?"

"You guessed it, Jack... Bingo Lizard Straits. Word has it that there are three carriers, seven heavy cruisers, five lights, twenty-seven destroyers, nine motor torpedo boats, a paddle wheel tour boat and a geedunk truck in there."

"Should be able to line up a target or two, Admiral."

"That's the spirit, Jack. I knew you would say that."

"We'll make you proud, Admiral."

"I know you will. By the way, the entire crew of the tender, Damage Control School, base sickbay, base galley staff, barbers and command staff have volunteered to stay aboard tonight and help your E-3s load torpedoes, stores, sea print films and trading material to use for barter with aboriginal simple people... and people in France... and paint the entire boat. By the way, Jack, how are Alice and the boys?"

"Well, Admiral, she was so despondent after the last assignment you gave us on the USS *Happyfish*, that she drowned Billy and little Teddy and shot herself."

"Hmmm, sorry to hear about that. Doris wanted to get her pineapple upside-down cake recipe."

Who writes the dialog for those gahdam things? Better yet, who does the Navy give them for 'technical advice'? Mary Poppins?

I don't know a damn thing about nukes. I figure all the movies made about them are the gospel truth. John Wynn told me, "Nukes never lie."

Standing Lookout

At times, there is no job in the entire world better than standing lookout on a diesel boat. Well, maybe it's number two behind being a professional beach comber on a little-known Pacific island paradise where the female inhabitants all have perky boobs and run around buck nekkit. Unfortunately, the latter never showed up on *Requin's* Watch, Quarter and Station Bill.

Adriane Stuke and I were professional lookouts. We both held Doctor of Relative Bearing degrees. With extensive postgraduate work in floating debris... Crap in the water... Oil slick identification, and 'What'n th' hell is that shit?' identification.

The only qualifications you needed to apprentice for the lookout position were (A) a pair of eyes, (B) the ability to drink liquid synthetic lizard dooky that the night cook passed off as coffee, (C) a 55gal., self-venting bladder assembly, (D) a minimal understanding of the '360 degrees in a circle' concept, (E) the ability to put up with boring conversation for hours at a time, and (F) personal plumbing fixtures big enough to locate in cold weather and that could extend farther than skivvies, dungarees and two pair of foul weather pants. The latter being by far the most important qualification.

In wintertime, being a lookout was, beyond any doubt, the most miserable, thankless, wet, cold, and never-ending job in the armed forces. I have never been so gahdam cold in my entire life. At times, your heart actually pumped ice slush through your veins.

When you're cold, miserable, laminated in more clothing than Tutankhamen, with a watch cap pulled down over your eyes, wearing mitts the size of boxing gloves...staring through fifty-year-old binoculars with lens scratches that look like ice skaters school figures...it's a damn wonder we didn't hit something. There were times I was so damn cold that I actually envied Joan of Arc. If you've never taken a leak in sub-zero weather in the sail of a pitching diesel boat, you've missed one of the all-time defining moments of life.

I know why the Titanic clipped the berg, the lookouts were cold. They were stamping their feet and rubbing their hands...the 7x50 binoculars were fogged up....

"Hey, Jack, why are we up here? We've got radar. Hell, when they invented radar, it should have let us off the hook. See them airplanes flying around? You don't see silly sonuvabitches sitting on each wing looking for stuff. Jeezus, it's cold."

"Bill, take a look...bows' on...zero, zero, zero.... What's that?"

"Who gives a shit...?"

WHOMP!

You didn't have to be exceptionally bright to read running lights, figure the 'angle on the bow', recognize a steady bearing rate, report 'red over red' (you know, the old 'red over red, the captain is dead' thing), read channel buoys and pass contacts to the idiot doing the same thing you were doing on the other side of the bridge.

There were silly things that lookouts did to new officers, things like during night steaming where you just ran to charge batteries then return on station. We used to see the moon make a 360-degree trip around the horizon and knew that the helmsman was giving the new guy a merry-go-round ride. A waste of time, complete circle where the helmsman threw a loop in the wake and the new guy missed it.

Another little stupid 'welcome aboard' stunt was to call out, "I've got a Bee-One-R-Dee... Bearing one seven five... Position angle fifteen degrees."

Bee-One-R-Dee...bird.

Or a 'Bravo-Two-Echo-Romeo casing'. Translation, 'beer can'. Both a highly worn out 'ha ha', but fun if you could toss the OD in the trick bag.

Coffee always tasted best on the bridge. You had to be good to climb the ladder in those high bridge fiberglass sails with three or four cups of hot coffee balanced between your left arm and your chest. If you never did it, you have no idea what I just said. If you did, you have the complete picture. And no cigarette with coffee ever tasted better than the one you got in the messdecks after being relieved by some other poor miserable bastard.

In SUBRON SIX, we used the old white Pyrex cups. When you finished your coffee, you put the empty cup in the void behind the radar mast. When the watch was over, each guy put a couple in his foul weather jacket pockets and took them down.

If they called up with, "Bridge...Conn. How many men on the bridge?"

You knew what was coming next, so you grabbed the damn things and tossed 'em over the side. The CO didn't want to dive the boat with half a dozen Pyrex cups doin' the mambo in his fiberglass sail, and he didn't want a lookout to fill his foul weather jacket with the fool things, busting one on the way down and arriving in the conn with a three-inch Pyrex shard sticking in a lung.

I'm going to laugh like hell if they display artifacts removed from the Titanic and five or six white Pyrex cups turn up... There's gotta be a few thousand of the damn things roaming around on the floor of the North Atlantic.

I'm proud of my 'years in the shears'. Met a lot of fine people and saw a lot of interesting stuff. I'm sure nukes have robot video cameras, satellite observation or some kind of electronic Seeing-Eye-Dog device. Damn shame, it was those experiences that casehardened your balls.

I can remember balmy summer nights, light breezes, full moon with reflection running all the way to the horizon. Boat running 'full on four' slicing along at twenty-plus knots, and bottlenose dolphins leaping around in the bow wave. Leaving phosphorescent tracks...water rising up the tank tops, slamming through the limber holes then falling away aft. Diesel exhaust drifting low over the screw guards to disappear in wake spray and the night. The luminescent glow of the stern light marking our passing. At times you could see the trailing edge of the flag aft of the sail, and when you couldn't see it, you heard it snapping in the wind. At times you could pick out the wing lights of aircraft heading to and from Europe. Once in a while, you got merchant surface contacts. Port and starboard lookouts speculate on what that tanker crew had for evening chow earlier and how much the sonuvabitches are making a month.

On rare occasions, you got a seagoing ocean liner. Skipper radios captain of the liner and tells him of our presence. Tells him we are a U.S. submarine. Asks him if he holds us on radar and can identify our lights. Both skippers agree that if the passengers see a surfaced submarine, we'll become an attractive curiosity, drawing too many folks to the rail. We darken the ship, turning off the running and navigation lights.

There we are, laying to in the dark. Beautiful ship passes...people doing triple flip-flops into the pool…women in dresses dancing with guys in their civvies class 'A's'…band music drifts across the water.

"Hey, Stuke…"

"Yeah, Dex…"

"You know what I want to do someday?"

"No telling…"

"I want to ride one of those big sonuvabitches. Have some pink-nippled blonde fluff up my pillow, scratch my back and sing me to sleep. Set my clock for midnight…get up, go down to the grand salon for champagne, shrimp and lobster tail. Take in the sights of nude swimming hour. Make a couple of bets at the O-3 level dog track. Catch a massage and sauna. Call the 'send me something soft and blonde to sleep with' steward and hit the rack."

"Armstrong…"

"Yeah?"

"You on dope?"

"Nah…just dreaming in Cinemascope. The price is the same, might as well go wide screen."

"Why don't you guys knock off the horsecrap? One of you drop down and rig out the running lights."

The Big Picture was Never in Focus

Me 'n Stuke worked for Arleigh Burke. We had no gahdam idea what middle management between the skipper and '31-knot' Burke looked like or did for a living. And frankly, didn't give a damn. Admiral Burke was a bluejackets' leader. He was a meat eater, absolute King of the Jungle. He was an action man in a world of 'all talk, no do'. Burke was the kind of individual who would hunt Bengal Tigers with a .22 and drag the dead ones home. No shark would ever eat Arleigh Burke out of reciprocal professional courtesy. Every mother in America could have no finer wish for a son than wishing he would grow up to be just like Admiral Burke. The heart of a lion packaged in a kind gentleman who understood leadership from A to Z. As far as we were concerned, the squadron staff on *Orion* were shore duty personnel nine to five useless overhead. Outside of constantly losing our pay records and hauling us up to sickbay and poking hypodermic needles in our butts, they never seemed to be doing a whole helluva lot that contributed to the 'Big Picture'.

Officers talked a lot about 'Big Picture' stuff. I think they dabbled in it on a kind of 'nibble around the edges' basis. If there actually was a big picture, it never reached Hogan's Alley on *Requin*, that's for damn sure.

Speaking of big pictures, in 1959 if some clown had come down Pier 22 with a forty-foot-high photograph of the squadron commander on a sixty-foot pole, every E-3 topside would have said, "Who'n the hell is that?"

If he turned up on *What's My Line?* and we could have won two weeks at The Waldorf-Astoria with the Playmate of the Month, we would have still been stumped.

We knew he existed because several times a day some peabrain on Mother Onion blew a damn bos'un's pipe and announced to a world that could have cared less, "Subron Six arriving..."

But as far as we were concerned, he lived in the same nether world with the Tooth Fairy, Santa Claus and the Easter Bunny, and other folks we had never met or seen.

We saw the Atlantic Sub Force Commander, Vice Admiral Elton W. Grenfell. He was a good egg. I don't have any idea what he did for a living other than sit around in a room jam-packed with four-stripers and think up weird stuff for submarines to do.

One time he came aboard. We knew he was a big cheese because the COB underwent a fast-pop religious conversion and had us take down all the nekkit lady pictures, including the really great one of Janet Pilgrim in a lace nightie taped to the inside of the after battery head stall door. An admiral wearing three stars could have had a mammoth attack of the green apple quick step and still there was no way in hell he would have gone in there. Didn't matter, the COB always won. We spent half the damn night turning to converting our clubhouse for seagoing idiots into the best imitation of a Sunday school we could come up with. And for the rest of my enlistment, whenever I was parked on the aft head in the after battery, I stared at the little pieces of tape that outlined where Janet and those 44 DDs peeked through that flimsy white nightie had been and cussed Sublant.

The admiral came. He had porked up a little since he tied knots in Tojo's tail and he had a sizeable pack of staff toadies nipping at his heels. They formed us up in dress canvas, including mess cooks. The standard drill, two ranks of stationary seagull crap targets aft of the sail. The admiral gave us the mandatory 'You men are doing one helluva job for The U.S. Navy' speech...the one where the duty mess cook always has that 'What in the hell is he talking about?' look on his face.

"Yes, sir, gentlemen...wish I could fill you in on the big picture and you would understand how vital your individual contribution is."

Always big picture bullshit. It always came by one-ton loads. After the speech, the admiral came down each row of bluejackets and spoke personally to each one of us. He didn't have to do that, but he did. You can't help liking an old smokeboat bastard who makes you feel like he really wants to shake your hand and say something to you.

He asked me if there was anything I would like to ask him.

"Yes, sir. Is there any way you could work it so the geedunk truck would take Canadian money?"

He looked at me like I had three heads and a tail and moved on. The COB looked at me like I was a total idiot.

Here was my idea: At some time or another, every ship based in Norfolk put into Halifax. When it shoved off, the bunk locker drawers were loaded with left over Canadian money roaming around in them. Canadian money in '59 had a par value greater than U.S. money. In sizable amounts, the difference could add up. If the roach coach took Canadian money, it would substantially increase sales because it would be the only place thousands of bluejackets could dump the stuff. The Navy mobile canteen folks would get a boatload of it and make out like Chinese bandits on large amount exchanges.

But if an E-3 thinks it up, it's gotta be stupid. Not 'big picture'.

After we broke quarters, the COB came over shook his head and said, "Next time the admiral shows up, I'm locking you in the paint locker."

I'm not sure that at nineteen the big picture matters much. Political alliances change. National identities change. Enemies come and go. You figure all that out much later in life. 'Big pictures' never remain the same. Maybe they aren't really 'big pictures' at all, just snapshots of moments in time.

At nineteen, the Russians were the bad guys. All targets were designated 'Ivan'. They rode boats hauling ordinance destined to be parked in your backyard. They were vermin and you were the Orkin man. A boatload of Russian boat sailors flooded at 350 feet was a cause for celebration. That was that much less ordinance available for package delivery.

Recent events make clear that somewhere around 1960 the big picture got refocused and things changed. But we were young. Didn't have time for anything resembling the big picture. Beer was a buck a pitcher at Bell's. Slim Jims were a dime and on a lonely night, if you were lucky, a barmaid would take you home for a hot shower and a late breakfast.

'Thirty-one-knot Burke' could shuffle around the big picture and we had complete faith in him.

If Grenfell ever showed up again, I had some other ideas to bounce off him, like moving the damn dumpsters on the pier closer to the nest, robotic chipping hammers and paint scrapers, and putting a couple of gals on the boats as backscratcher's mates.

Women on submarines. At nineteen...single...and a long way from home. Not a bad idea.

Once Upon a Time

One of the benefits of growing old is the gift of time. Time to look back and revisit your collective 'Life Experiences'.

For old smokeboat sailors, that means time to shuffle through memories of pissing against the wind in faded soft dungarees, frayed raghats and zinc chromate-spattered broghans. You can close your eyes and be transported back to a time when men wore acid-eaten uniforms, breathed air worse than the primate house at a poorly managed zoo, whittled mold and rot off food of advanced age being reclaimed by the gods of putrification, and surgically carving off the stuff and eating it. You survived and built up an immunity that could handle leprosy, lockjaw and cobra bites. We survived. Submarine duty was rough.

Many of us 'hotsacked'. For those of you who missed that life experience, 'hotsacking' was sharing sleeping arrangements (to put it in easily understood terms). A system that required lads at the entry level of the undersea service profession to crawl onto a sweat-soaked flashpad just vacated by another bottom-feeding shipmate. Lads of today's modern, technically advanced undersea service would find it damn near impossible to imagine a day when lads who hadn't showered in weeks climbed a tier of racks sharing sock aroma on a par with three-day-old roadkill with his bunkmates. A time when raghats communally shared blankets that looked like hobo camp hand-me-downs.

It was a time when the common denominator of the naval supply system was the cockroach, with the longevity of Jack LaLanne. Cockroaches that could deflect claw-hammered blows and could reach rodeo entry size.

In the late fifties, the submarines built in the twilight years of World War II were rapidly approaching an advanced-age comatose state. The Navy quit making many of the replacement parts for these seagoing antiques so we cannibalized the boats in line heading to the scrap yard. It was like harvesting organs from a dead Rockette to keep the chorus line going. After decommissioning, the old boats would have electricians and machinists crawling all over them with shopping lists and wrenches.

Memory is a wonderful God-given gift. There were sunrises and sunsets, rolling seas, visits to exotic places, and ladies with loose panty elastic and no AIDS. There were consumable combustibles on a par with the liquids that propel hardware to outer space.

It was a time when the world's population loved the American submariner. Boatsailors in port meant good times, hell-raising and calling in the nightshift at the local brewery. It was a time when the United States Navy had no recruitment problems, paid no incentive money and had to kiss no butts to entice grown men into accepting their manly obligation to their nation. Men signed up for undersea service, motivated by patriotic obligation, a sense of history and adventure, and to follow the gallant submariners who rode the boats against the Japanese empire. We wanted to wear the distinctive insignia universally recognized as the symbol of the most successful and demanding submarine service on earth.

We were proud. We had a right to be. We were accepted as the down line fraternity brothers of the courageous men who put Hirohito's monkey band all over the floor of the Pacific. We rode their boats, ate at their mess tables, slept in their bunks and plugged the ever-increasing leaks in the hulls they left us. We patted the same barmaids' butts they had patted when they were far younger and half as wide. We carved our boats' names and hull numbers on gin mill tables in places that would give Methodist ministers cardiac arrest.

We danced with the devil's mistress and all her naughty daughters. We were young, testosterone-driven American bluejackets, and let's face it, every girl in every port establishment around the globe both recognized and appreciated the meaning of a pair of Dolphins over a jumper pocket. Many of these ladies were willing to share smiles and body warmth with the members of America's undersea service.

It was a time when the snapping of American colors in the ports of the world stood for liberation from tyranny and the American sailor in his distinctive uniform and happy-go-lucky manner stood for John Wayne principles and a universally recognized sense of decency, high ideals and uncompromising values.

It was in every sense of the term, 'A great time to be an American sailor'.

There were few prohibitions. They were looked upon as simply unnecessary. It was a time when 'family values' were taught at family dinner tables, at schools, the nation's playing fields, scout troops, Sunday school or other institutions of worship. We were a good people and we knew it.

We plowed the world's oceans, guarding her sea-lanes and making her secure for the traffic of international commerce. But at eighteen, let's face it, we never thought much about the noble aspect of what we were doing. Crews looked forward to the next liberty port, the next run, homeport visits, what the boat was having for evening chow, the evening movie after chow, or which barmaids were working at Bell's that evening. We were young, invincible and had our whole lives ahead of us. Without being aware of it, we were learning leadership, acceptance of responsibility and teamwork in the finest classroom in the world: A United States submarine.

It was a simpler time. Lack of complexity left us with clear-cut objectives and the 'bad guys' were sharply defined. We knew who they were, where they were, and that we had the means, will and ability to send them all off to hell in a fiery package deal. We were the 'good guys' and wore 'white hats', literally.

What we lacked in crew comfort, technological advancements and publicity, we made up for in continuity, stability and love of our boats and squadrons. We were a band of brothers and have remained so for over half a century.

Since we weren't riding what the present-day submariner would call 'true submersibles', we got sunrises and sunsets at sea. The sting of wind-blown saltwater on our faces…the roll and pitch of heavy weather swells and the screech of seabirds. I can't imagine sea duty devoid of contact with these wonders. To me, they are a very real part of being a true mariner.

I'm glad I served in an era of signal lights, flag messaging, navigation calculation, Marines manning the gates, locker clubs, working girls, hitchhiking in uniform, quartermasters, torpedomen and gunner's mates, sea store smokes, hotsacking, hydraulic-oil-laced coffee, lousy midrats, jackassing fish from the skids to the tubes, one and two-way trash dumping, plywood dog shacks, messy piers… A time when the Chief of the Boat could turn up at morning quarters wearing a Mexican sombrero and Jeezus sandals. When every E-3 in the subforce knew what paint scrapers, chipping hammers and wire brushes were for. When JGs with a pencil were the most dangerous things in the Navy. When the Navy mobile canteen truck was called the 'roach coach' and sold geedunk and pogey bait. When the breakfast of champions was a pitcher of Blue Ribbon, four Slim Jims, a pack of Beer Nuts, a hard-boiled egg, and a game of eight-ball.

It was a time when if you told a cook you didn't eat Spam or creamed chipped beef, everybody laughed and you went away hungry. And if you cussed a mess cook, you could find toenail clippings in your salad.

It was a time when, if you saw a boatsailor with more than four ship's patches on his foul-weather jacket, he was at least fifty years old and a lifer. A time when skippers wore hydraulic-oil-stained steaming hats and carried a wad of binocular wipes in their shirt pockets. In those days, old barnacle-encrusted chiefs had more body fat than a Hell's Angel, smoked big, fat, lousy-smelling cigars or 'chawed plug', and came with a sewer digger's vocabulary.

It was a time where heterosexuals got married to members of the opposite sex or patronized 'working girls', and non-heterosexuals went into the Air Force... Or Peace Corps.

It was a good time... For some of us, the best time we would ever have. There was a certain satisfaction to be found in serving one's country without the nation you so dearly loved having to promise you enlistment bonuses, big whopping education benefits, feather-bed shore duty, or an 'A' school with a sauna and color TV.

Our generation visited cemeteries where legends of World War II undersea service were issued their grass blankets after receiving their pine peacoats and orders to some old hull number moored at the big silver pier in the sky. We were family. Our common heritage made us brothers.

There came a point where we drew a line through our names on the Watch, Quarter and Station Bill, told our shipmates we would see them in hell, shook hands with the COB, paid back the slush fund, told the skipper 'goodbye', and picked up a disbursing chit and your DD-214. We went up on Hampton Boulevard, bought a couple of rounds at Bell's, kissed the barmaids, gave Thelma a hug, then went out to spend the rest of our lives wishing we could hear, "Single up all lines..." just one more time.

Why Submarines Are Better Than Women
Mike Hemming

1. A submarine will kill you quickly. A woman takes her time.

2. Submarines like it done at all angles.

3. Submarines can be turned on easily anytime.

4. A submarine doesn't mind if you smoke, drink, tell dirty jokes or cuss.

5. A submarine does not object to being rigged for dive.

6. Submarines come with manuals.

7. A submarine is built for going down.

8. A submarine once down is quite willing to say there as long as you want.

9. Submarines are always in trim.

10. You can dive a submarine any time of the month.

11. Submarines don't whine unless something is really wrong.

12. Submarines don't care how many other subs you have sailed on.

13. Submarines don't come with in-laws.

14. When sailing, you and your submarine arrive at the same time.

15. Submarines don't mind if you look at other submarines or if you buy books and magazines about them.

16. Submarines don't complain if you sleep somewhere else.

17. Submarines don't mind if you stop off for a few beers on the way back.

18. It's okay to tie up a submarine.

19. You can leave your submarine but it will never leave you.

20. You get paid extra money for riding on a submarine.

21. Submarines don't mind if you sit up all night eating and talking loud with your friends.

22. Submarines always smell the same all month long.

23. A submarine doesn't get mad if you fart, belch or scratch in public.

Der Drache ist Todt
The Dragon is Dead
Grant 'TarJack' Swinbourne

Kiel Harbour, Northern Germany, September, 1941 04:00

The early morning air had a crisp autumnal bite to it that made Leutnant-zur-See Gunther Kruse hunch his shoulders as he lifted his scarf to cover his neck. As he strode down the pier he watched the harbour starting to come to life. Fishing boats returning or just setting out skimmed across the small waves blown up by this cold north wind. Only seven or eight knots he thought to himself, just a gentle zephyr really.

He knew out beyond the protection of the Baltic and on the North Sea that cruel mistress that had claimed so many of his friends lives, even at his tender twenty-three years, would show its full fury to him and his crew. Possibly during this patrol…

He knew what it was like out there. He'd already served four patrols in *U-48* with some close scrapes with the British destroyers. They had still managed to sink over 32,000 tons in their last two patrols.

He wondered how he would fare as the skipper of the boat. How would his men respond when the time came? Would they be able to withstand the shattering noise and the shaking of the depth charges? Would he be able to bring them all and the boat back safely and in one piece? Would the boat be able to handle the punishment that he knew only too well could be dished out?

He let his mind turn to the last patrol when they were attacked on the surface by an RAF fighter. They had only just escaped, and three of the watch crew had lost their lives as they struggled to get into the conning tower hatch. He shuddered at the thought and returned to the present.

Kruse was in a hurry to get back aboard *U-56*, he was returning from collecting his sealed orders from the harbour kommandant's office. His second-in-command, Walter Hoss, was already aboard making sure that their final preparations were concluded and they would be ready to sail as soon as Kruse came aboard.

As he came around the corner of a warehouse his command and pride and joy came into view. Over two hundred metres of grey steel, the slim upper casing glistened darkly in the moonlight. The conning tower fat with two tiers for the flak guns mounted at the rear of the platform that served as the boat's bridge, on which he and his watch crew would be spending a significant portion of this patrol. A thing of warlike beauty, *U-56* was his first command and his chance to prove himself to his superiors.

As he came aboard Oberfahnrich Schulman and the watch crew formed up on the bridge and piped him aboard. He stopped at the foot of the gangway turned sharply to the stern of the boat and saluted the flag flown on the short staff at the far end of the casing.

He climbed aboard and was greeted by his second-in-command, Hoss, who was just emerging from the hatch. "Everything is in readiness, Herr Kaleun!" he reported.

"Then let us begin," replied Kruse.

Hoss and Schulman rapped off their orders to the deck crew and engine room. The roar of the twin diesels starting up obliterated anything but the loudest petty officer's roar. The deck crew efficiently slipped the mooring and *U-56* was finally underway. Kruse's breast swelled with pride. This was for him one of his proudest moments. From the time that he had first been to sea whilst a naval cadet in the Kriegsmarine, he had wanted a command of his own, and finally it was his.

The U-boat slid quietly out of Kiel harbour and turned towards the Kiel Canal entrance. As she slid quietly under the cover of the predawn darkness, past the Graf Zeppelin, Germany's unfinished aircraft carrier and the other ships of various shapes and sizes, he thought back to the night before, carousing with his comrades, some from his old boat congratulating him on his promotion.

I shouldn't have had that last bottle of champagne. The thought pounded through his head like the rivet guns already at work in the yards on the far side of the harbour. The rolling of the boat on the slight swell didn't make him feel any better. Next time it would be different. No drinking before leaving. For many of his men though it would be a very different story. He knew every inch of the boat and how the workmen had put it together. He was lucky enough to have been given command during the last stages of refit so he'd personally supervised some of the work, particularly on the interior fit out.

His hand caressed the cold steel edge of the conning tower. He knew the steel must withstand what he asked of it. He turned to his watchoffizer, Martin Schulman, a young oberfahnrich from Hamburg. "The weather forecast looks good so far."

"Ja," replied Schulman. *A man of few words*, thought Kruse.

Schulman had been a last-minute replacement for this trip and Kruse hadn't had a lot of time to get to know him. *Ah well, plenty of time for that on this patrol*, he mused. He thought of his wife, Ingrid, and the cottage they shared in a small village near Bremen. She would be sleeping now and possibly dreaming of him. What did she really think of him and his involvement in this war? Only seeing him every few months when he could get away on leave. That would be even harder now that he had a command.

His last leave was only three weeks ago and already he yearned to be with her. He remembered their last night together, Ingrid's warm body next to his in their bed, the guttering candle on the nightstand the only light in the room. She would never say it out loud to him, never tell him not to go back. But he felt it. He knew she hated him for it. He thought about how they had parted with a kiss, almost chaste, after a silent breakfast, the issue of his imminent departure hanging over the table like a spectre.

The submarine stayed at eight knots as they followed their escort out of the harbour sea walls, between the submarine nets and minefields that guarded the harbour entrance. An hour or so later, Kruse went below as he ordered the boat to ahead standard. They could cruise at this speed (around thirteen to fourteen knots) for almost three weeks, slowing only when they dived for regular hydrophone checks to listen for the telltale sound of a merchant's screws churning through the water. If they slowed to nine knots they could cruise for over a month almost non-stop.

The distinctive smell was the first thing that greeted Kruse as he descended the ladder into the control room. Although the boat was newly painted and refurbished inside and out, the smell of the forty men that were to live in it had already started to mingle with the diesel and oil stench that would live in his nostrils and on his skin forever. One note that stood out was the unmistakable waft of wurst, ham and eggs cooking.

Walter Britzling, a leading seaman who had sailed with Kruse on three occasions, was already cooking breakfast in the miniscule galley behind Willie's cabin. Not that it was much of a cabin, more an alcove with a curtain, but certainly more privacy than was afforded to the rest of the crew. They had the indignity of sharing bunks, with the on-duty crew waking the off-duty crew and climbing into their warm racks. Each crewman was given a rack about five-and-a-half feet long and about three feet wide, suspended from the inner pressure hull of the boat by a chain at either end. There were three bunks layered on top of one another with about two feet between each bunk.

Being fully provisioned for a four-week patrol meant that the crew also had to share their bunk space with any food or other supplies that couldn't be stowed elsewhere. On the first day of a new patrol almost every inch of spare space the boat had was crammed with tinned and fresh food, ranging from eggs, hams, sausages and fruit, to spare torpedoes and fuel drums, crowding the already cramped compartments. Even some of their drinking-water tanks had been filled with fuel to provide that little bit of extra range.

Directly opposite Kruse's bunk were the hydrophone and radio rooms. There two specialists would sit for most of the patrol listening on the radio for messages that might indicate a convoy was close enough to move on for an attack. The hydrophones were checked every few hours by diving to around twenty metres and running at slow speed and making a few turns of the boat to make sure that they covered all angles of the compass. Kruse's bunk was also a seat for three when dining; a foldout table was set up in the narrow passage between his bunk and the radio room.

As he and two of the other officers sat down at the folding table to eat, conversation moved to what their orders were. Kruse couldn't open the orders for another two days. Their first patrol grid was BE83, one of the hundreds of patrol grids the BdU had divided the North Sea, Atlantic Ocean and Mediterranean Sea into. Admiral Karl Doenitz, the supreme commander of the Ubootewaffe knew exactly where each U-boat was at any given moment. Kruse presumed their mission would be to intercept convoy traffic moving across the Atlantic and into either the St George's or the English Channel, the strategic choke points for convoys. Occasionally they would expect to see southbound convoys carrying supplies for the British effort in North Africa.

Because of the heavy defences and minefields in the English Channel, they would have to go the long way around the north of Scotland, down the western coast of Ireland until they reached their patrol grid some one and a half weeks from now.

As the boat entered the Kiel Canal, Kruse watched the moonlight playing on the calmer waters of that narrow ditch that crossed the southern end of the Danish Peninsular and his thoughts fell to the work he and his crew had already done in preparation. His crew was a mixture of first timers and more experienced men. His job on the trip out would be to whip them into shape as a crew. This would mean constant exercising, when not actually in contact with the enemy, so they would be ready.

One of the biggest tasks was to keep the men occupied. Boredom frequently set in during the long patrols through what seemed like empty seas waiting for some contact with the enemy.

The trip through the Canal that morning was uneventful and there were no delays to speak of waiting for the locks to fill and empty. As they left the last lock, the first rays of the rising sun were warming their faces. The breeze was much as it was when they had left if not a little fresher. Martin Goetz, the lookout on the forward port watch, spotted a glint of sunlight off metal low on the horizon.

As they drew nearer they could hear the strain of two high-revving diesel engines pushing the E-boat at around twenty-five knots. "One of ours," said Schulman. The E-boat was the surface equivalent of the U-boat. Fast, light, and highly manoeuvrable, they were responsible for coastal patrol and raiding around the German-controlled North Sea and Atlantic coasts. This one was painted with what looked like a tiger-striped camouflage pattern that broke up her lines in an effort to confuse enemy gunners. Kruse ordered a signal to the patrol boat asking if they had seen any traffic in the area. The response "All clear" came back and *U-56* continued on her way in a northwesterly direction.

"Sail spotted," cried young Goetz, pointing to a few points off the port bow. All sets of eyes on the bridge swung to try to catch sight of the new vessel only two hours since their encounter with the E-boat. They were still in the German Bight and it was likely that this would be another German vessel, but they still had to be careful. Reports of commando raids being dropped on the French and Dutch coasts meant they were alert from the moment they left the quayside.

"Looks like a schooner rig," said Goetz. "She's flying the swastika." *This kid has good eyes,* thought Kruse. Sure enough, after another hour the schooner hove into plain view. A privately owned sailing boat was a rarity now that the war was in full swing. As they came to within a few hundred metres the master of the schooner hailed them.

"I saw a small British steamer about a day ago, moving slowly to the west," he shouted through an old-fashioned megaphone.

"What are you doing out here?" hailed Kruse.

"We're taking a small cargo from Zeebrugge to Hamburg."

"What cargo?"

"I can't tell you that."

"We will have to board you, then." Kruse was a little curious as well as frustrated by the reply. Stopping this boat even though it was German was necessary. It could be an enemy ruse and they could also be carrying contraband even if they were German.

"Sure, come aboard and have a look for yourselves."

"Take a boat and five men over and check that everything is in order," he ordered Hoss.

An hour later the schooner and the U-boat parted. "They were carrying a couple of crates of Dutch masters," said Hoss. The artworks had been 'confiscated' by some official in Zeebrugge who didn't want to have them carried home over land or by air for some reason. "He's taking a risk, though if he's seeing British shipping, isn't he?"

"I'd say so," said Kruse. "Unless the Englander was way off course."

"Well, based on what he told me the vessel should be somewhere around here by now. Do you think we could catch him?"

"Well, if he doesn't change speed we should be on him by the end of tomorrow and he should still be out of range of most of the English patrols." Hunched over the small plotting table in the control room the CO and his number two worked out the course and speed they would need to catch this freighter.

"Set course, 248 degrees, rudder fifteen degrees, speed fourteen knots," Kruse ordered. The helmsman and engineer responded and the chase was on.

By mid-afternoon the next day they had moved well into the North Sea and it was Schulman this time who spotted the smoke on the horizon. "Full ahead on electrics," ordered Kruse. "Give me seven metres." He wanted to close to gun range before the merchantman spotted them.

The sub settled low in the water so that only the conning tower was cutting through the waves. At their lowered profile they would be harder to spot, but they had to move slower. The diesels couldn't operate when the hull was immersed. The exhausts and intakes would ship seawater and flood the engines, so the electric engines were started. Only a dim whine in the engine compartment gave away the fact that they were underway.

U-56 sported an 88mm Krupp gun with an effective range of over five kilometres, but firing from the unstable platform of the U-boat casing was another matter entirely. To truly be effective he needed to get as close as five hundred metres.

Schulman called up the gun crew. They raced up the ladders and unlocked the gun from its cruising position, removed the waterproof coverings, and opened the hatch for the ammunition locker. A human chain extended from the locker to the hatch and to the gun to pass ammunition as fast as it could be fired.

The schooner's lack of response to the fast-approaching U-boat puzzled the *U-56* officers and crew. The merchant hadn't appeared to have spotted them and was gently cruising along at five knots, just another tramp steamer going about its business.

"Do you think it could be a Q-ship?"

"Hard to say, Herr Kaleun. If it is a trap, you'd think you would see a bit of activity on the deck, but I don't think they've got lookouts posted. I can't see any movement in the masts at all."

The submarine closed with the merchant and when they got to within three kilometres, Kruse gave the order to prepare to commence firing. "Aim for the waterline!" shouted Schulman.

"Aim for the waterline…I'd like to be able to hold this tub still enough to just hit him from this range," grumbled Unteroffizer Kurt Frank under his breath. As gun layer, it was his job to keep the gun levelled at the target despite the rolling deck beneath them. This was no easy task even in these calm seas and he also had to constantly adjust the range as they closed on their victim.

"Feuer!" yelled Kruse and the boat shook with the reverberation of the shot.

"They've seen us now," joked Hoss as the 88mm shells screamed towards their target.

"Ja, keep a look out for aircraft and have the flak crew stand by." Kruse didn't want to bring too many more men above decks in case they had to crash dive. Every man on the bridge or casing meant precious seconds during which they could be hit with shells or bombs.

The merchant started to turn away from *U-56*, but it was to no avail. After only twelve shots, most of which were hits, the small cargo ship had launched its lifeboats and was well on its way to becoming a smoking hulk.

"Good shooting, boys," said Kruse. "Let's get out of here." He ordered a turn of speed to put some distance between them and the evidence of their attack. He didn't want to risk slowing down to question the crew of the stricken freighter while he was within RAF range from England. He was proud of his men, they had performed flawlessly. He had the radioman, Franz Dorner, send off a contact report advising of the sinking. It was only about 1,200 tons, but it was still their first sinking as a crew. Their celebration was brief, with congratulations being passed around and a friendly atmosphere pervading the submarine. Even a small victory was a victory, and each one was welcomed.

The boat turned back onto its northwesterly course straight towards the Orkney Islands north of Scotland. It would take them another two days to reach the waters surrounding these remote islands and then another four days to get to their allotted patrol area, as long as everything went to plan.

Over this time Kruse exercised the crew mercilessly. Diving stations were called every three to four hours. They had to dive to do a regular hydrophone check anyway, but sometimes Kruse ordered crash dives that took them as deep as sixty metres. The drills would save their lives if they were attacked by patrolling aircraft. Luckily, so far none had been spotted and more importantly, none had spotted them. The crew went through endless damage control drills until they satisfied their captain's wish for perfection, loading and unloading the torpedoes from their tubes in mock attacks against the empty sea.

⊕ ⊕ ⊕

"Smoke spotted, range about eight kilometres, bearing 201!"

This time it was Joachim Tripp who raised the alarm. Kruse had been sleeping in his alcove, it was 04:15 and about another two hours before dawn. He grabbed his leather coat and struggled into it just before launching himself through the narrow circular open hatchway to the control room. Kruse glanced at the dials on his way to the ladder that led through the conning tower to the bridge.

Hoss was on watch on the bridge. "Morning, Herr Kaleun, looks like we have a bigger fish this time."

"Let's see." Kruse raised his binoculars and scanned the inky black horizon. In this light it was hard to tell sea from sky, but Hoss's arm pumping up and down along the bearing line gave Kruse a good start. He could see a dim glow as if from a hidden fire and a slight smudge of smoke against the blackness. Now that his eyes had adjusted he could also just make out a bow wave. Judging from the distance between the glow and the wave, it was certainly a much bigger ship than their first victim.

Again, Kruse ordered his gun crew to stand by. He felt a primitive greed to conserve torpedoes, even at the risk of being shelled. They made a bow-on approach to the ship so that they would be harder to see against the rising sun. "It's either an ore carrier or a tanker from the configuration of its masts." The recognition books weren't much help until they could get close enough to identify her properly.

This time the lookouts on the ammo carrier were doing their job and the crew on *U-56* saw her starting to take evasive action. "Surface and muster the gun crew on the casing," Kruse ordered. Efficiently, the gun crew emerged from the hatch and went through their routine of readying the gun for action.

At 1500 metres, they started firing, and this time it was only a few shots before a massive purple and black explosion ripped though the forward half of the ship. "She's carrying ammunition!" screamed Hoss over the searing holocaust. Secondary explosions ripped the dawn sky. A massive explosion toppled the aftermast and the ship's funnel, then tore the stern of the ammo carrier apart and she quickly sank. No boats or survivors were spotted.

"She had to be at least 5500 tons," said Hoss, grimly elated.

The gun crew in particular were very pleased with themselves and proclaimed that torpedoes were no longer needed, and that they could throw them all overboard next trip and load up with high explosive 88mm rounds.

Kruse had Radioman Bruno Wendt put some of his jazz collection onto the gramophone. The scratchy tones of Louis Armstrong started belting out *When the Saints Come Marching In*. Music drifted out of the radio room through the submarine and could faintly be heard through the conning tower to the bridge. Most of his collection was banned in Nazi Germany, but he had managed to conceal the records in the record sleeves of good German military marches. "What would Onkel Karl Doenitz think of his crews listening to *untermensch* music?" Kruse mused as he drifted into an exhausted slumber. He didn't care, because he knew his men enjoyed his collection as much as he did.

The next morning, just as the first rays of sunlight crept over the horizon, the lookout cried, "Ship spotted!" It was another small freighter and heading almost right for their submarine.

"Dive! Dive!" shouted Kruse. "Take her down to thirty metres and level out at ahead one-third."

U-56 responded to the downward pressure on her forward dive fins and dipped her bow beneath the gentle swell just as the last of the lookouts cinched the upper hatch closed. The sound of rushing water filling her tanks, drowning out the hubbub below.

"When she closes to two kilometres, we will blow ballast at flank speed, get the gun crew on deck and start the fun," explained Kruse to his officers in the control room. "Schulman, I need your gun crew on their toes for this one. We won't have a lot of time once we are on the surface before the target radios our position in. Aim for the bridge and then the waterline. I want this one sunk in under ten minutes."

"Jawhol, Herr Kaleun." Schulman turned on his heel and shouted for the gun crew to assemble in passageway to the control room while Kruse ordered the forward torpedo tubes checked, locked, and flooded. Their audacious plan would require critical timing. The crew of the freighter would not be expecting an enemy submarine to surface next to them in daylight. Kruse hoped the tactic would give his men a few seconds head start over the lookouts on the merchant.

The tension was as thick as the humidity in the crowded control room. Condensation formed on almost every exposed metal surface and the maps and charts spread out on the plotting table stuck together.

"Bring her up to periscope depth," said Kruse as they closed the gap.

Moments later, "Periscope depth, Kaleun."

"Ahead slow." Kruse had already climbed the ladder to the conning tower where the periscope was located. The periscope slid almost silently upward until it broke the surface. Kruse waited until the glass cleared then swung the scope around to scan the horizon for threats. Nothing other than the freighter was in sight and it had closed a little faster than he had expected. No matter, it made the wait a little shorter. He kept tracking the freighter for another thirty-five seconds, just enough time for him to update his speed estimate.

The scope slid back, "Ahead flank," he cried.

"Jawhol." The message was relayed to the motor room. The motors whined and sparked in protest as they were wound to full power, pushing the great brass propellers faster and faster through the water. It would make a small wake on the surface but would scarcely be noticed by the freighter in the cold morning light.

Kruse counted off the seconds in his head and marked the chart he was using to plot the attack with a grease pencil as the two vessels closed the gap between each other. The gun crew crowded the passageway, nervous and excited.

"Blow ballast, ahead flank. Ten degrees starboard rudder," he roared.

They broke the surface at seven knots, the ballast blowing increasing their top speed slightly as *U-56* broached and crashed off the top of a wave into the trough following it.

The lookout on the bridge of the freighter spilled the fresh mug of tea that had just been brought to him and called out the alarm to the officer on duty. "Bloody hell!" he exclaimed, "Submarine, there on the starboard quarter, sir!"

The duty officer called down for the captain. "We need to make sure it's not one of ours before we radio it in," he said. "They may need our help."

On *U-56*, the gun crew raced to their positions and removed the waterproof covers as quickly as they ever had. "Range 1600 metres, bearing 285 degrees."

The gun layer already had the first shell in the breach and was locking it off.

"They don't look bloody helpless!" cried the freighter's lookout, pointing out the gun being trained on them. Before the officer could look up, the bridge erupted in a ball of orange and red flame as the 88mm shell ripped through the wheelhouse door, killing everyone in it in seconds.

"Good shot, Franck!" applauded Schulman.

He's actually enjoying this, thought Kruse as the gunners reset their weapon.

The next six shots saw the small freighter slowly sliding nose first into the depths below. "Another 2500 tons with just the gun," exclaimed Franck. "You can chuck those eels overboard now, we really have gotten the hang of this."

"Just wait till we see some bigger game, then your pop gun can be the anchor we use to tie up with next time we're in harbour," said Torpedoman Grodl.

U-56 resumed its course and sailed on for the next three days, diving for hydrophone soundings every few hours, without a sighting and luckily without being sighted.

During the afternoon sound check, Henner, the sonar man on duty, called out, "Herr Kaleun, I think I may have something, bearing 56 degrees. It's quite faint, but I think it is multiple screws so it must be a far distance off at the moment. Too far to pick out revolutions anyway."

Kruse joined his soundman in the cramped hydrophone room and took the headset.

"Good work, Henner. I think we may have a convoy. Chief, change course to track this contact for a while. Ahead slow."

As he listened, Henner gave a running commentary on the sound contact's bearings, while Schulman and Kruse plotted them on the chart as the minutes ticked by. "Looks like they are closing on our position," Schulman murmured.

"Ja, I think our torpedomen may be in luck this time," said Kruse with a crooked smile.

The sounds of individual ships were distinguishable as they drew closer. "I'm getting at least ten separate contacts," Henner said. "Maybe more...at least two escorts." Henner could now hear the higher-pitched sounds made by the smaller ships' propellers as they made short fast bursts to circle around their charges.

"Given their speed and the fact that they will cross our path in about an hour, I'm going to attempt a daylight submerged attack," Kruse told Schulman. "If we have to we can fall back and follow them until dark, then attack on the surface later tonight. Have the tubes loaded with the electric torpedoes to avoid the wakes of the steamers; set with impact pistols." Forward in the torpedo compartment, they groaned as they knew the next hour would be a race against time to unload two of the tubes loaded with steam-powered torpedoes and reload them with electrics.

"Take her down to sixty metres. Speed four knots."

"Sixty metres, speed four knots, jawohl, Herr Kaleun."

The helmsmen adjusted their trim and plane controls and the blades at the bow and stern of the boat bit into deeper water, angling the nose of the submarine towards the ocean floor. The twin electric engines sent a steady hum through the boat, pushing it towards its deadly rendezvous.

"Thirty metres...forty metres...fifty metres, levelling off."

"Slow ahead, silent routine," came the order as up and down the length of the boat the men stopped all activity, and some even perceptibly changed their breathing in an attempt to remain quiet. A few quiet creaks from the outer hull came, reminding everyone aboard of the hundreds of pounds per square inch of pressure that threatened to crush them inside their tin can.

The thrumming swish of the convoy's propellers was loud in Henner's earphones as the British ships edged closer and closer. "They seem to be turning towards us, Herr Kaleun. I'd say there are at least six heavy ships, and at least ten other smaller ones."

"Steer 250 degrees, ten degrees rudder. I want our nose pointed at them so they don't get too much of a sniff."

Louder and louder now came the cacophony of screw noises. The only other sound was Henner's whispered reports on the bearing changes as the convoy started to spread out around *U-56*, closing unknowingly with their destiny.

"Bring her up slowly to periscope depth now, chief."

The men sweated, silently waiting at their posts for the call to action. Confident, but also wary. The previous engagements had gone smoothly with no hazard to the boat. That could change in an instant. One bad break and they could be the victims. In the forward torpedo room, all tubes loaded and ready for launch, the men traded silent glances, hoping that they would have their chance to show their mettle to the gunners, who were lying on their bunks aft.

Slowly, almost imperceptibly the submarine rose as small amounts of air were pumped into the trim tanks. As the first of the escorts passed noisily overhead, the boat was levelled skilfully at periscope depth. The noise from the convoy was clearly heard throughout the boat without the need for hydrophones now as they came to within 1500 metres of the surface ships.

"Watch your depth, chief," murmured Kruse. "We don't want to tip them off now, do we? Raise the attack scope; keep her steady and slow to one knot. I don't want to show a wake." Kruse climbed into the conning tower as the scope slid noiselessly upward. As it broke the surface he swivelled for a quick view to check for the positions of the escorts and to see if there were any aircraft around. At this depth, in the right conditions, they would be able to see the sub as a shape under the water and would immediately attack.

"Down scope. No aircraft, but I see six destroyers," he noted to Schulman, who was at the plotting table updating the marks as the positions of the closest ships were called down to him.

They were 1200 metres away and closing on the starboard outside line of ships in the convoy. "Target; light freighter, 1100 metres speed now six knots, angle on bow about forty-five degrees," he called down to Schulman. "Second target in line astern about 400 metres behind the first same speed and course."

Schulman made some calculations and started turning the dials on the torpedo data computer to set up the shot.

"Flood the tubes and open tube doors."

"Update the solution, Schulman, and plot it on the map," Kruse ordered.

"Solution ready."

"Scope up. Closer…closer…torpedo los. Left rudder, ten degrees, come to course 235 degrees." The boat jumped as the first eel was expelled by compressed air. The electric torpedo's engine started twenty metres from the U-boat, pushing it to a speed of thirty-five knots.

Kruse swung the periscope to the next target. "Range 850 metres, speed still six knots. We'll use three torpedoes on this one. She's big enough. Eight thousand tons I estimate."

"Tube two, los. Tube three…los. Down scope, tube four, *los*. Dive to eighty metres, make four knots." Kruse nimbly slid down the ladder to the control room and cinched the conning tower hatch closed as the crew responded and the boat began its descent.

The first torpedo announced itself with a deep boom. Heads turned, grinning, eyes flashing, but none of the crew made a sound.

The other three torpedoes streaked toward their target. The second and third were hits, ripping huge holes in amidships of the merchant, but the first narrowly skimmed across her bows and ran on for another forty seconds until it hit another ship two ranks further on. This one suffered a fate similar to the others.

All three ships were now shipping water as other ships swerved to avoid colliding. The escorts now swung into action to wreak revenge for the sinkings, racing from their picket positions to start searching for the unseen raider below.

U-56 was now closing down for silent running as she slowed her descent and waited for the retribution they knew was coming. The pinging of the ASDIC had already started as the nearest escort turned and began an active search for the submarine. It was like a dance of two blindfolded people in a dark warehouse, one silent and trying to avoid capture, the other calling out and listening for the echoes. The pings echoed against the hull of *U-56* and some of the less experienced sailors winced as the noise kept on and on.

"Take her down to 120 metres," breathed Kruse and down dipped the nose of the submarine.

"Weak contact on bearing 320, sir."

The thrumming of the first destroyer became a thrashing noise, drowning out even the sharp reverberations from the ASDIC.

Wasserbombe.

The overhead splashes were clearly heard as the drum-like depth charges were rolled off the stern of the destroyer. "Ahead flank, starboard ten degrees, down to 150 metres," ordered Kruse in an effort to manoeuvre out from under the hunter above.

As they sank, the depth charges tumbled and spun, trailing bubbles as they dropped deeper and deeper. At a hundred metres, they began to go off. Henner threw his earphones onto the table next to him as his eardrums felt like they had been hit with a hammer.

Silent routine.

Again and again the ringing from the ASDIC hammered at them. Again and again they were subjected to the tense waiting for the final blow that would allow the crushing pressure at this depth to seal their collective fates. Again and again they were surprised to find that they were still alive, but constantly hoping that the next one would be further off than the last. Just a little farther would bring their hopes up.

Each time the weapons were heard dropping from their hosts above, Kruse made course and depth changes in the hope that he could shake off his pursuers. Keeping the boat at two knots and making no more high-speed dashes he wanted to make sure the boat was as silent as possible with all the noises being made by the explosions. The crew could hear the thrashing of the propellers of a circling destroyer overhead.

"Flank speed, hold the rudder at zero degrees. Down fore and aft planes fifteen degrees, take her down, Chief," Kruse ordered. The boat shuddered as the gearing was engaged to push her faster and downward, but there wasn't enough time for the boat to get to full speed before the first explosion ripped through the water, the pressure wave hitting *U-56* like a sledgehammer hitting a garbage can.

Three more explosions, each closer than the first, came in quick succession, rocking the hull of the U-boat. The second broke loose some of the lagging around the pipes in the control room, raining down white dust on everything like snow. A fitting broke loose and water streamed around the bushings. "Damage control on that pipe. Make it fast."

A rating jumped over with a wrench in his hands and started to tighten down the bolts holding the flange. Throughout the boat there were groans as crewmen who had lost their footing started to feel their injuries. Luckily, most were minor, although one had broken his forearm, the bone jutting through the skin. The shock had been so great he hadn't noticed the pain

until he looked down at his shattered arm.

"Medic!" screamed the petty officer in charge of the compartment as the man started to howl in pain and panic. Several more explosions close by wracked the boat. The vibrations this time caused several small cracks through which water from the trim tanks was now being pushed by more than 600 tons per square inch of pressure. Pressure which was increasing as the submarine nosed down towards the sea floor thousands of metres below.

Kruse calmly ordered the sub to be levelled off. He knew that wouldn't be enough with a steady flow of water into two of the boat's compartments aft. They may continue to drop. The effect of the depth charges was now only psychological, ripping almost overhead as the U-boat had dived below them.

The men were now ankle-deep as they worked to stem the flow. Mattresses and wooden beams and wedges were manhandled into position in an attempt to seal the leaks. Kruse marked an hour of hard labour with cold water inching slowly up their legs. The pumps working their hardest to push the water back into the tanks, all while *U-56* crept slowly deeper and deeper into the ink black Atlantic, flirting with crush depth.

"Two hundred metres," said Hoss, sweat streaming down his forehead.

"Two hundred and ten...twenty.... Slowing... slowing, two hundred and twenty-five... Two hundred and twenty metres and *rising*." An electric thrill rippled through the crew. They were still alive.

The rumble of depth charges was now the only sound as exhausted crewmen collapsed onto the still saturated deck plates. Despite the noise of their exertions it appeared the Tommies had lost them for now. Intermittent pings rang out, distant, searching, but to no avail.

"One hundred and ninety metres, Herr Kaleun," reported Hoss.

"Hold her steady at hundred and ninety. Maintain silent routine," came the calm reply. "Well done, gentlemen. The practice drills have made a difference."

"We'll have to put the pumps on soon to get rid of this water," advised Hoss in a low whisper.

"Give it an hour and we'll see where these bastards on the surface are. Then we can think about home comforts like a dry floor."

The submarine crept along at only two knots. A slow crawl at best but enough to manoeuvre with. The sound of depth charges receded as the destroyers had clearly lost them or had gone chasing another echo they thought more promising. After another hour they couldn't be heard. However, Schmidt on the hydrophones could still hear the receding convoy and the sounds of the still-circling escorts listening for any sign that the U-boat was still close by.

"How long till sunset?" Kruse asked.

"Sunset is due in another two hours, Kaleun," Schulman replied. Kruse did some mental calculations after checking the battery charge and the CO^2 meter.

"Bring her up slowly to sixty metres." He didn't want to risk being down too deep for too long. He knew that the damage to his hull might be fatal at these depths and he wanted to make sure he had a good chance of getting his men to the surface if there was a failure of the pressure hull. "Maintain that depth for another half hour before starting the pumps and securing from silent routine."

The sound of the convoy and the escorts was fading even in Schmidt's headset. As they reached sixty metres the pumps were restarted. The noise was jarring as the water was sucked down through the bilge pumps and expelled into the trim tanks. A cautious sweep of the periscope and *U-56* was once again a surface craft.

A steady rain hit the deck plating as the twin diesels pushed *U-56* along at thirteen knots. The chase was on to try to regain contact with the convoy and to hopefully get in front so they could set up their next attack later that night. The gloom of dusk was made even darker still as the cloudbank moved in to smother the setting sun as the rain started to fall.

Below the crew off watch was lying exhausted in their bunks recovering from the afternoon's excitement. In the engine compartment and the control room, the on-watch crew continued their jury-rigged repairs.

"We'll have a hard time picking up the convoy in this weather.," said Hoss.

"I know, but we must press on," replied Kruse. "They can't be more than fifty kilometres away at the speed they were going."

"But what if they have changed course?"

"We should know by morning, then." Kruse sank into his own thoughts. Thoughts of home and hearth. His dog Gunther resting its head on the rug beside the fire. Ingrid making dinner while he read the newspaper. It wasn't so long ago since his last leave in November of last year. Not so long ago. "I'm going below to get some sleep. Call me if we pick them up," he said.

Kruse reached the bottom rung of the conning tower ladder, called for a coffee and told Heschler in the radio room to put on a record. He slumped onto his bunk and rubbed his eyes. The coffee came. Ersatz, with the strong odour of toasted acorns. It was an acidic brew and while he still hadn't gotten used to it, with enough sugar it could be made just about palatable.

In the ensuing hours he snatched sleep, updated his logbooks, and made three trips to the navigation table to check on their progress. During his third visit that he decided it was time to dive and listen for the convoy. The watch crew clambered dripping wet down through the hatches and cinched the hatches closed.

"Take her down, Chief. Thirty metres, please, and slow ahead when we get there."

"Planes down fifteen degrees, fill forward tanks."

A minute or so later Kruse was in the hydrophone room hanging over Henner's shoulder as he turned the wheel that directed the hydrophone to a compass bearing.

"Nothing," said Henner as he completed his first sweep. He was straining his ears as he bent low over the console, his head twisted to one side as he listened for the disturbance caused by the convoy through the staccato noise from the bad weather above.

Minutes ticked by monotonously. Kruse drew deeply on his cigarette as he leaned against the open hatch that led to the control room. "We can't have lost them, even in this weather, Hoss."

"I know, I know. But if they changed course a few degrees we could be getting further away rather than closer."

"I'm not going to let the chance for another attack slip away that easily. Surface the boat, all ahead full, lookouts to the bridge."

"I think it's dryer at thirty metres!" laughed Hoss once they had climbed to the bridge.

"Ja, it's a foul night," replied Kruse. "I just hope we don't run into the escorts before we can see them."

⊕   ⊕   ⊕

The driving rain and wind whipped up a spray that made the watch crew's search so much harder in the darkness. After four and a half hours of pounding through the waves and rain, Kruse said, "I want to do another sound check, Hoss, let's dive to thirty metres again and drop to slow speed."

As they reached the required depth, the watch crew were warming themselves in the forward mess with coffee and hot chocolate. Kruse, still dripping, hovered near Henner as he tried to regain contact with the convoy.

"There they are... bearing 347 degrees," murmured Henner. "Contact getting stronger."

A minute or so passed as Kruse and Henner shared the experience of hearing the scratching grow into an arrhythmic thrumming with a low rumbling roar as a background.

"It looks like they're steady in the bearing so we're on a parallel course, sir."

"Hoss, let's see if we can get a plot on them. What range do you think, Henner?"

"They would have to be at around twenty kilometres out; with the noise of the weather though it's hard to be accurate."

Kruse joined Hoss at the plotting table as they worked out the distances, fuel consumption, and angles with slide rules. "If we run at full speed for another three hours, we should be able to get around the outside escorts and in front of them," said Hoss.

"I'd like to get a look at them if we can, but I don't think we can risk it in this weather, unless we get a break in the rain. Surface the boat, ahead full, get the forward tubes ready."

"Nice night for it, hey, Skipper?" shouted Hoss as the vibration of the radio aerial cables started their familiar moan.

"Ja, but I'd rather be in the Petit Porc having a schnapps in front of Mme Dumar's fire." They both settled into their silent vigil, struggling to see any further than a few hundred metres as curtains of rain flew almost horizontally across their view.

Rolf Lunders crawled up through the conning tower hatch and tugged at Kruse's raincoat. "Radio message from BdU, Herr Kaleun. Oberfahn

Sohsrich has decoded it already."

From: BDU

To: U 59

GOOD WORK ON CVY. KEEP IN CONTACT AND WAIT FOR FRIENDS TO FORM UP IN BF3945. EXPECT 2 AT APPROX. 0545.

Kruse glanced at his watch. That was another four-and-a-half hours away.

U-56 approached the planned rendezvous. The radio traffic increased as two other U-boats closed in on the ambush area. Kruse and his crew had kept updating BdU and the other boats of the convoy's progress and the three boats had arrayed themselves across a twelve-kilometre stretch, lying like lions in wait for their prey. Each boat taking turns in diving for fifteen minutes and updating their colleagues on their hydrophone soundings via radio when they surfaced. They now had a pretty good picture of the speed and position of the lead ships in the convoy and their only problem now was the foul weather that was growing worse.

The night was now black as pitch as heavy curtains of rain swept across the ocean surface, cutting visibility to less than one kilometre at times. The watch crews were being rested on hourly rotations as the boats rolled and pitched in the heavy swell. Large waves were rolling in succession across the path of the convoy, forming calm patches in the lee of the huge rollers.

"In this muck we should be able to slide between the columns without being seen," Hoss said.

"I know but with so many targets and with three boats, we will still get a result from outside. I'm not sure I want to be on the inside when the torpedoes begin flying left and right," said Kruse quietly. "Henner, where are they now?"

"Four kilometres and closing, sir!"

"Let's surface and try to get closer on the diesels before running decks awash for the attack."

"The decks will be awash anyway in this muck," muttered Hoss.

Kruse and his watch crew clambered into the cold wet space at the top of the conning tower, which was barely big enough for them to stand shoulder to shoulder. They strained to see through the murky blackness as they crept closer to their prey.

Visibility had improved slightly as the rain was now just a faint drizzle, but that meant visibility for the convoy's lookouts had improved also. Kruse mulled over the decision to attack on the surface and tried to calm himself with the knowledge that they could always dive to escape.

"Drop to seven metres depth," he ordered. "Start electric motors."

This would make the U-boat almost silent as they approached the convoy. The waves were now not breaking at all but gradually rising and spilling over the lip of the conning tower sending sheets of freezing water down the hatch into the tower and control room below.

"Ahead one-third, crew to action stations. Quietly now," Kruse rasped down the hatch to Schulman.

The watch crew were now huddled together straining to see into the darkness as their boat crested the rollers, collectively holding their breath. Dimly ahead of them, the grey-white bow wave of a ship started to take form.

Normally a sighting would have generated an excited cry from the lookouts, but this time the mood was subdued and anxious as they struggled to make out what shape was above the wave. Hoss caught sight of a steeply raked bow now curving above the wave as the U-boat slid into a trough between two rolling giants, leaving them blind as the ship disappeared into a trough of its own. "I think it could be an escort!" he said, his voice almost a whisper.

"Hold on," Kruse said, as they rose on the swell of the following wave. "This is going to get interesting." As the boat wallowed over the crest of the roller they could see the outline of the British corvette only a scant kilometre or so in front of them. There didn't appear to be any reaction on the escort. Their luck was holding so far.

Hoss held his breath as they drew closer to the escort with no response. The tension on the watchtower was palpable as the U-boat and the corvette passed within 800 metres of one another and still no reaction from the British ship. The corvette was now slipping astern. The anxious watch crew stared as the grey shape slid into the gloom behind them.

"I never want to get that close again!" whispered Hoss.

"Nor I, my friend," murmured Kruse. "Keep your eyes open, men. Aft watch, keep lookout for that corvette. Forward watch, we should be seeing the freighters soon." *Was the corvette the forward or port side escort?* If their calculations were correct it must be the forward escort, but if they were wrong. The entire convoy would slip past them to port while they groped around in the dark. Submerging for another sound check was out of the question when they were this close with the poor visibility on their side.

As the minutes ticked by and the wind continued to drop, they could see a rainsquall slowly drifting away behind them as if a curtain was lifted and suddenly they could see a full two kilometres in front of the boat.

"Ship spotted!" Matrosengefreiter Schect was pointing to starboard as the bow wave and then the high prow of a large freighter emerged from the gloom.

"Hard to port, another knot of speed, please, Chief." Kruse's order was relayed down the voice pipe to the control room and agonizingly the bow of the submarine started to turn as the freighter rumbled unseeingly towards them, the bow wave rocking the stern of their boat as they escaped disaster by a few metres. The grey black bulk of the freighter passed behind them as they straightened on their attack run on the next row of ships, which were now dim shapes growing in definition before their eyes.

"Prepare a stern shot calculation for that freighter. Speed six knots, bearing now 168 degrees, angle on bow 182 degrees." Below, Schulman worked on the complex math to provide the firing solution and then entered the details into the torpedo data computer and adjusted the dials as he watched the stopwatch.

Inside the cramped stern torpedo room the crew worked feverishly to flood the tubes. "Firing solution ready, Herr Kaleun!"

"Start calculations for the forward tubes, two targets, a tanker and an Empire class freighter. Both on the same heading as the stern shot and angle on bow of seventy-five degrees and closing."

"Gyro settings ready on tube five, Herr Kaleun. Firing solution ready on forward targets."

"Flood all tubes." Kruse bent over the UZO scope, watching the first of his three targets creeping towards the crosshair. First the prow then the mast followed by the superstructure "Fire tubes one and three!"

He quickly swung the UZO to align with the target to his stern. The process repeating with an unwavering inevitability. Now swinging back to the third target and again the cool efficient orders were being rapped out and responded to with mechanical accuracy. His stopwatch now running he called for the bridge to be cleared for diving.

"Watch crew below, clear the bridge for diving. Dive planes fifteen degrees, starboard rudder ten degrees, ahead one-third; continue on electrics." These last orders were delivered as he dropped off the ladder into the control room, having cinched the hatches behind him. "Chief take her down to eighty metres.

"Eighty metres, jawohl."

"Torpedoes running on course, Herr Kaleun," called Henner from the hydrophone room. As their depth increased they could hear the noise from the convoy's engines and propellers clearly through their hull. The hand on Kruse's stopwatch continued on its way, bringing the destruction of the freighters closer with every tick.

The dull thuds came, making the hull vibrate with the sound waves travelling across the kilometre or so of water that now separated the U-boat from its first victim. "Torpedo impact!" Schulman called noting the time in the combat log.

Three more impacts came in quick succession as the boat continued its dive. Then another and another still. "One of the other U-boats must have fired not long after we did," said Schulman.

"Well done, gentlemen," said Kruse. If we make it through this, we'll have a celebration. He thought about the crates of Beck's beer under his bunk saved for just such an occasion.

In the forward torpedo room, Muller made a rude gesture at Frank as grins were exchanged and backs slapped. There was no conversation as they now listened to the tortured screams of the metal of their targets breaking up. The U-boat hull gave off a soft groan as the pressure increased as if in sympathy. The crew knew only too well that it could be their turn very soon if one of the escorts found them.

Two more distant crumps were heard as the torpedoes fired by the third U-boat hit home. The only other noise was the chief reading off the depth as they settled at eighty metres.

"Two torpedoes still running, sir. I think they've missed."

"Thanks, Henner." Kruse glanced at Schulman, who nodded, confirming the misses. "Ahead slow, change course to 355 degrees, Chief. Maintain silent routine."

"Jawohl, Herr Kaleun. Steer to course 355 degrees."

The thrumming of the convoy's engines grew fainter as it slipped astern. Suddenly the sharp crack of a depth charge broke the stillness. Everyone's nerves were on edge, but they recognized that the sound came from a distance and wasn't near enough to harm them.

"I think they've found one of the others," Hoss said quietly as more explosions were heard.

"Let's hope they don't get a lucky hit with one of those eggs," Kruse replied. "Increase our speed to three knots."

The boat cruised on in near silence. Only the now increasingly faint sounds of depth charges disturbed their quiet progress. Eventually all external sound died away. All that could be heard were the familiar sounds of water hissing past the outer hull and the dripping of water inside the compartments from the condensation in the thick and stale air.

After another hour of steady running. Henner reported that the convoy was on its way out of earshot. Kruse ordered the boat back to periscope depth, lifting the noise restriction and starting the celebration by dragging the crates of Beck's out from under his bunk.

"Set planes fifteen degrees up angle," rapped the chief, a wide grin spreading across his rugged, old-before-its-time face.

The crew took ten-minute turns up on the bridge as the thin grey dawn broke. The seas had moderated and rather than the steep swell that had made the attack so hazardous, there were smooth lines of far more gentle rollers.

A flurry of radio traffic between BdU and the boats informed them that all three boats had survived and between them had sunk over 75,000 tons of enemy shipping. Kruse read the totals out to the crew as they crowded the compartments at either end of the control room. Cheers and glances showing how pleased they were to have survived the night unscathed.

⊕   ⊕   ⊕

"I think I'm picking something up."

Kruse leapt up and crossed to the hydrophone station. He grabbed a spare earphone and listened intently as Henner trained the hydrophone receiver towards what sounded to Kruse like a faint scratching somewhere off in the distance. Another day and most of a night had passed as *U-56* paced around the expected convoy track.

"Bearing 307 degrees," murmured Henner. "Contact getting stronger. I think we're closing range fast."

The sounds that Henner reported were like no other ship he could recall hearing. To him it sounded like there was a cable wrapped around the propeller, there was a distinct clang each revolution.

"Put us on their bearing, Chief. Surface the boat, give me full power. We've got about an hour before dawn to get in position," Kruse shouted. "Lookouts to the bridge, have Goetz take this watch, we need his good eyes."

U-56 burst to the surface and the twin diesels roared to life. Kruse, Hoss, and the lookouts moved into position on the bridge. The eastern sky was faintly purple with the approaching dawn. The U-boat raced across the whitecaps, occasionally showering the bridge with spray.

"They should be in that direction," Kruse said. "Do you see anything, Goetz?"

Goetz nodded his head. "Smoke, quite a lot, Herr Kaleun."

The sky gradually lightened to reveal a long, broad column of smoke marking the horizon. As *U-56* closed the gap, they could make out a solitary vessel.

"Look at that smoke pour out of her, she's smoking like a dragon," Hoss said, drawing laughter from the men.

"She's rather low in the water aft, too. I think we've found a straggler from our convoy. Let's help her put out that fire, men. We'll attack on the surface; gun crew to the control room, forward torpedo room ready for action!" cried Kruse when he and Schulman had calculated the timing of the attack. Frank and his gun crew crowded into the control room as the boat rose slowly to the surface.

They were about five kilometres away from their target as they sped through the gentle swell. Quickly and efficiently the gun crew readied the 88mm gun for action, removing the waterproof covers and sliding the first round into the breech.

Hoss and Kruse were alone in the conning tower as they discussed the plan of attack. The target was still making only three knots and hadn't deviated from its course.

As they closed to within two thousand metres, they could see the British ship was alive with activity, men scrambling fore and aft, manning hoses while alarms screeched in the early morning air. Parts of the cargo were smouldering but no flames were visible.

"Fire!" shouted Kruse.

But the first shot came from their target. First a flash followed quickly by the crack of a twenty-five-pounder gun mounted somewhere on the bow of the British ship. The shell screamed as it passed overhead and in response the German eighty-eight roared as their first shot was fired. The crew aboard the British ship sprang into coordinated action, lowering facades and sidings to reveal two other guns, which then opened fire.

"*Alarm!* Gun crew below! Leave the waterproofing! Get ready to dive!" Kruse wasn't going to lose his command on his first patrol. He shoved Hoss towards the conning tower hatch as the gun crew scrambled up the ladder one at a time, Hoss had relayed the order to dive and the crew, responsive as ever, were already filling the ballast tanks and angling the planes downwards.

In the bridge Kruse watched as his gun crew slithered down the hatch the last man, Frank, slipping as he reached the top of the external ladder from the deck to the bridge. His chin hit the bridge deck hard and his mouth filled with blood as he bit into his tongue.

Kruse dived to reach for Frank's hand, but it was too late. Already the swirling waters were lapping at the top of the ladder where Frank had been, his body slid astern. More shells screamed overhead as Kruse got to his feet and struggled to help Frank over the railing.

In the control room below anxious moments passed. Already water was streaming down and unless the hatches were closed the whole crew would perish in seconds. Hoss started to climb the ladder, his heart heavy with thoughts of the loss of two good men. A commander that he had served with for a few short months in this stinking war.

Frank's large frame tumbled headlong though the hatch. Then Kruse's feet, legs and trunk followed as sheets of water now hit him full in the face. Kruse hauled the hatch closed and cinched it locked just in time as one of the twenty-five-pounder shells hit the lip of the conning tower and ricocheted into the sea.

Both Kruse and Frank lay heaving for breath as they recovered from their ordeal. Hoss shouted into the control room below, "Medic to the conning tower. Emergency dive, take her down to one hundred metres."

Then the pinging started.

"It must be a Q-ship!" cried Hoss as the incline of the deck increased.

The first salvo of depth charges came a few minutes later as they heard the thrashing of the Q-ship's propellers driving it overhead. "It looks like we kicked a *sleeping* dragon this time," Kruse said.

After a half-hour of sporadic depth charging, Kruse decided to go on the attack. "Periscope depth, set one forward tube to magnetic, the others to impact pistol. Set the stern tube to magnetic. Schulman, what's the draft of that vessel?" Kruse ordered.

Schulman rapped out the depth almost immediately. He still had the book open at the page they had for the surface attack and had anticipated the request.

"Set the magnetics at seven metres, the impact pistols at five metres."

As usual, his commands were carried out efficiently as the submarine smoothly glided upwards. Kruse knew they were making more noise than he would have liked, but it couldn't be helped if they were going to sink this menace.

"Boat at thirteen metres, Herr Kaleun," Hoss called out as Kruse climbed the ladder into the conning tower. He glanced down at the blood on the floor and for some reason thought of Ingrid and had an odd moment imagining that at the very same time she was thinking of him.

"Up periscope." He waited until his view cleared of the water. The Q-ship was about 1200 metres way on their starboard bow. "Come starboard fifteen degrees. Flood all tubes and open the outer doors," he called. He could see the ship making a smart turn to port, the smoke greatly diminished now, bringing it in line to cross his bows in a few minutes. He called out the course and speed to Schulman, who worked furiously to calculate the solution. The bows of the Q-ship touched the center crosshair of the scope and he waited until the forecastle of the ship had glided by before issuing the order to fire.

He slid down the ladder ordering the scope to be lowered and then ordered increased speed and depth. In the hydrophone station Henner followed the sounds of the torpedoes as they raced toward their target.

The magnetic torpedo passed under the bow of the Q-ship. The resulting explosion rocked the ship to port, buckling steel plates near the keel and starting a deluge into the lower decks. As the ship rocked back towards the collapsing column of water thrown up by the explosion, the second torpedo struck the hull at an odd angle and the impact pistol didn't ignite. The torpedo slowed a little as it scraped under the hull and then increased speed as it cleared the ship and continued into the dark water beyond. The third and fourth torpedo missed the target entirely. "One hit. Three misses, Herr Kaleun," reported Henner.

"Damn!" Kruse glared at Schulman, who looked away.

The Q-ship completed its turn and started dropping depth charges.

"Set speed ahead standard, rudder full port turn. Bring her back up to periscope depth, Chief. We'll see if we can hit him with our stern tube."

Kruse climbed back into the conning tower and raised the scope to see the Q-ship listing to starboard and slowing. As the stern of *U-56* swung towards the stricken ship, Kruse calmly called out the range and speed to Schulman.

"Rudder zero degrees. Open the stern tube door," Kruse called. "*Los!*"

The boat shook slightly as the torpedo was released from its restraint. Kruse kept the scope up and ordered a hard starboard turn, aware the wake of the scope was being tracked by the enemy.

Fateful seconds ticked by as they made their turn, and Kruse kept the scope glued to the target. He watched in awe as a giant plume of water leapt into the air, settling back down over the decks and superstructure of the Q-ship. A second later the crack and boom of the explosion could be felt rather than heard through the hull of the submarine. A secondary but far larger explosion ripped the Q-ship into two parts, showering the surrounding water with burning pieces of debris.

"Der drache ist todt," Kruse whispered.

Submarine sims, what a true child of the modern entertainment world. My first subsim experience was back in 1985, in the barracks on the sub base at Groton. I remember looking for other "targets" once the canned targets were all gone and the sub ran off the game map! The next was Aces of the Deep and Silent Hunter, (in between was a sea tour on the *USS Sunfish*). Silent Hunter hooked me. Being an active submariner made it a bit easier to swap sea stories and relate to the vets I spoke to. I found Subsim Review, a place where I was welcome to donate the 20+ years of knowledge and facts I had accumulated. Over the past three years I retired from active duty, bought a house, and entered the civilian work force. To this day I dearly miss the boats & the friends I served with. SH3 and SH4 have been a real pleasure to experience, my hope for future subsims is that the realism and technical accuracy improve to a level unseen in even the best sims we currently have available to us.

Frank "Torpex" Kulick

Submarines from Containment to Preemption
by Capt. Zeb Alford (ret.), USN

President Truman announced the policy of 'Containment' in 1945. The greatest war the world had ever seen, World War II, was just over. Europe, Japan, and many other battlefields were devastated areas. Truman convinced the Congress that the country needed the Marshall Plan to rehabilitate Western Europe and Japan and keep them from falling behind the Iron Curtain of the Soviet Union. The Containment policy lasted forty-sevenyears until the Soviet Union went belly up in 1991.

In August 2002, eleven years after the demise of the Soviet Union and the policy of Containment, President George W. Bush announced a new policy – 'Preemption'. The president announced that in the future America would not wait to be attacked. We would engage any terrorist or country that supported terrorists anywhere in the world. In his speech at West Point, he amplified his policy of Preemption to protect America by adding that henceforth America would maintain an overwhelming military force to ensure that no nation would be encouraged to attack us in the future or attempt an arms race with us. Prime Minister Churchill said it best when he said to the House of Commons after America entered WWII: "America always does the right thing, but only after they have exhausted every other alternative."

The policy of Preemption was a dramatic shift from any previous ones in our history. During the nation's 229 years only three, I repeat, three, national policies have determined when we went to war, and what kind of military forces we had available to fight. The first of the three was Isolationism. George Washington announced this in his farewell address to the officers that fought with him in the Revolutionary War. He warned against entangling alliances with European nations. This policy lasted 170 years until the end of WWII.

Today, I want to connect the dots on how the shift from 'Containment' under President Truman to 'Preemption' under President George W. Bush affected the Navy's submarine force. I was a midshipman at the Naval Academy when President Truman announced his policy of Containment. I have personally served on or have visited all the submarines constructed from the 1940s until today. When I graduated from the Naval Academy in 1947, I knew I wanted to become a submariner. Our submarines during WWII were only 2% of the Navy, yet they sank 60% of the Japanese ships sunk. All of the Japanese merchant fleet was on the bottom of the ocean. There was a high cost for this outstanding record; 25% of our submarine force was lost. Fifty-two boats are still on patrol. This is the highest loss suffered by any part of our military in WWII.

In 1900 the Navy commissioned its first submarine, *USS Holland*. The sub was named for its inventor, John Holland. The first submarine had all the basic elements of our modern submarine except for its gasoline engine. It was small enough to fit in one of the missile tubes of today's strategic missile submarines. Over the next forty years the *Holland* evolved into the most effective diesel-electric submarine ever built--the fleet boat of WWII. The Navy built over 250 fleet boat submarines. With the record of sacrifice and success of the submarine force, at the end of the war there were no enemies, no targets, and no mission. The submariner force had to reinvent itself. It needed a strategy to support the policy of Containment.

The Soviets were building submarines as fast as they could by 1950. Soviet strategy was to be able to sink our carriers before we could launch an attack against the Soviet Union.

American submariners believe there are only two types of ships, submarines and targets. In 1950, Soviet submarines became our targets. We had no way to detect, tract, or sink a submerged submarine. This is the kind of challenge that submariners love.

The Containment policy became known as the 'Cold War'. It changed the submarine force dramatically. The Soviet Union and the U.S. were the only superpowers during the years 1945-1991. The military strategy to deter the Soviet Union became know as MAD (Mutual Assured Destruction). They thought they would bury us. Communism was the wave of the future; so they thought, but President Truman said nuts to Communism and the Soviet Union. The Communist party of America and the left wing of the Democratic Party have never forgiven America. Today we still see this same group of Americans giving aid and comfort to our enemies as they did in the late 'sixties during the Vietnam War. They were led by such great

Americans as Jane Fonda and her jolly band of brothers who I label 'Hate America Firsters'. They know what they are doing, and so do our enemies, and so do we.

After graduating from submarine school I served all of my twenty-six years of active duty as a submariner in the Cold War. What a war it was for a young submariner. After WWII, Admiral Rickover revolutionized submarine warfare by building the first nuclear-powered submarine. Before nuclear power, submarines ran on large batteries when submerged. They could run only thirty minutes at maximum speed, longer at slower speeds. In 1955 Commander Wilkinson sent the historic message "Underway on nuclear power" from the *USS Nautilus*. This submarine could run at full power submerged for two-and-a-half *years*. By 2003 our nuclear submarines had steamed some 126 million miles safely on nuclear power. Today, nuclear submarines can run the life of the submarine without refueling.

I became executive officer of the diesel submarine killer *Cavalla* in 1957 (yes, the same submarine now in Galveston). *Cavalla* is famous for sinking the large Japanese carrier *Shokaku* with three torpedo hits. This was on *Cavalla's* first war patrol in June 1944. She took over a hundred depth charges and earned the name 'Lucky Lady' and the Presidential Unit Citation.

Cavalla and three other fleet boats from WWII were converted to a submarine killer configuration in 1951-1955. They were all assigned to a new Submarine Development Group in New London, Connecticut. The conversion installed a large sonar array in the bow that could detect a Soviet submarine snorkeling thirty to seventy miles away. These four submarines were instrumental in developing the submarine killer tactics; the wire-guided torpedoes, the silencing of our submarines, and the sonars for the modern nuclear submarines.[1]

I was selected for the nuclear program by Adm. Rickover in 1958 and commissioned on the first nuclear submarine killer, *Tullibee*, as XO in 1960. After 'charm school' (serving on Adm. Rickover's staff), I commanded one of the first high-speed, single-screw, nuclear attack submarines—*Shark*. This submarine class was the first streamlined fast attack that could go twice as fast submerged as on the surface. We spent some time operating against our own carrier groups to show them what to expect from the Soviet nuclear submarines. The rest of the time we spent gathering intelligence about the Soviet submarines. Today, our modern nuclear attack

---

[1] Editor: See *Civilian Submariner* by Dr. Donald Ross, *2007 Submarine Almanac*

submarines are the silent escorts of our nuclear carrier battle groups with their ability to kill anything in or on the surface of the ocean, including other submarines and surface ships.

An example of how our submarines were used in the Cold War involved the *Shark's* operations during the Cuban missile crisis of October 1962. When the crisis started I had been at sea for about two months. My mission was to monitor Soviet submarine operations in the Barents Sea north of the Soviet Union. I reported a large number of subs leaving ports. I received orders to return to Norfolk, Virginia ASAP. We set a record crossing the Atlantic that probably stills stands. I arrived in Norfolk Navy Base on Friday. *Shark* had been submerged for seventy-two days, the longest submerged time of any submarine at that time. All the families were on the pier to meet us. Also, the admiral of the Atlantic Fleet submarine force, Admiral Joe Grenfill, was on the pier. He welcomed me back and wanted to know if any equipment was out of commission. I told him, no, but we were down to canned food and had used many spare parts to keep everything operating during this prolonged patrol. He then said he regretted to tell me this, but we had to overhaul all the torpedoes immediately. Also replace any conventional warheads with nuclear ones. My orders were to get underway Monday and sink any Soviet subs that got anywhere near our blockade line north of Cuba. All leave and liberty were cancelled. I told the admiral that we would be ready on Monday. By Monday the crisis was over. We went on leave. Containment worked, but just barely.

In 1963, after the loss of *USS Thresher*, I was ordered to the puzzle palace on the Potomac (the Pentagon). I was the only nuclear-trained submarine officer there. My marching orders from Adm. Rickover were to prepare the testimony to Congress and report accurately what caused the loss of *Thresher*, and what the Navy was doing to correct any problems. Admiral Red Ramage, one of my WWII submarine heroes and a Congressional Medal of Honor recipient, made the presentations to Congress. The Congress granted the Navy $2 billion to install the 'Sub Safe Program' on all nuclear submarines.

I commanded the 5th Polaris submarine to be built, *Sam Houston*. I never dreamed that I would command a capital ship of the Navy as a junior commander. Admiral Nimitz, the Texan who led the Pacific forces in WWII, knew better. He addressed his staff after the Japanese surrender in 1945 and said, "Before the war the battleship was the capital ship of the Navy. During the war the carrier became the capital ship. In the future, the submarine will become the capital ship of the Navy."

President Eisenhower instituted the Polaris program in 1956. This program commissioned forty-one Polaris ballistic missile submarines within fifteen years. The Polaris project was probably the largest and most successful industrial project in this nation since WWII. These submarines carried sixteen nuclear-armed missiles that could hit the Soviet Union with multiple nuclear warheads from 2500 miles away. Today it takes six years to build one nuclear attack submarine. During the Cuban Missile crisis in 1962 we had five Polaris submarines at sea targeting the Soviet Union. The Soviets had none at sea. The Cuban missile crisis was the nearest that our nation came to attacking the Soviet Union.

After leaving command of *Sam Houston* it was back to the Pentagon for me. There are some memories from my days in the Pentagon and the National War College at Fort McNair in Washington D.C. that I would like to share with you.

I was the Senior Naval Aide to the Under Secretary of the Navy (now Senator) John Warner from Virginia from 1969-1970, and graduated from the National War College in 1971. One of my heroes from my Navy days in the Pentagon was the CNO, Adm. Moorer. As you recall he later became the Chairman of the Joint Chiefs of Staff. One day his aide called and said he wanted to meet with the SecNav and UnderSec Nav. At that meeting he informed them that the British First Sea Lord had alerted him that the British would be pulling its Navy out of the Persian Gulf and out of all its Asian bases. If the Persian Gulf were to remain stable, we would have to replace the British. The British had been maintaining a naval force there since WWI, but they no longer had a budget to allow them to continue. He further stated that they were willing to lease us a deserted island called Diego Garcia in the Indian Ocean for ninety-nine years. Diego Garcia was about 1000 miles from the Persian Gulf and had a great harbor, but we would have to build an air and logistic base to support our Navy. I can still feel the embarrassment of not knowing where this island was. Secretary Chaffee (ex-governor of Rhode Island and later Senator) said he understood about the need for us to maintain the security of the Persian Gulf area, but we were pulling out of Vietnam, and with our budget cuts he didn't see how we could do it. Without batting an eye the CNO stated he was aware of that, and he thought he had a solution. He reminded the secretaries that when we launched men in space we could communicate with them until they passed the east coast of Africa, but we had no stations to reach them until they approached Australia. The astronauts had just landed on the moon, and he thought we could have NASA pay to establish a communications station on Diego Garcia. With that the secretary said *go*. Twenty-one years later when President George H. W. Bush commenced the Persian Gulf War of '91, the Marines landed with their equipment from

Diego Garcia, and the B-52 bombers flew from there. It is important to remember that we had admirals that could see twenty years ahead on how to protect this country. It is also important that we would fight to keep the oil flowing from the Persian Gulf. The bad guy was named Saddam Hussein. Since then we have invaded two of the most impossible countries to reach from the sea to take the war to the terrorist. It wouldn't have happened without Diego Garcia. Thank you, Admiral Moorer.

At the National War College we studied the power of various nations we might have to fight in the future. There were only three areas where we thought we might have to fight in the next twenty years. The Persian Gulf to protect the world oil supply, Southeast Asia to protect Japan, and any areas the Soviet Union might challenge us such as in Korea and Vietnam. We had seen that our strategy of Mutual Assured Destruction (MAD) had prevented an invasion of Europe by the Soviets. Our class was concerned about the number of nuclear weapons we and the Soviets were building. We were aware that it didn't take many missiles with multiple warheads to deter us or the Soviets from attacking each other. I wrote my thesis about stopping this escalation by making more of our strategic forces mobile. This could be done by rail or trucks on land or at sea by submarines and Navy surface ships. If you can't find them, you can't target them with a ballistic missile. This was not popular with my Air Force classmates since they were in the process of buying 1000 long-range ballistic missiles in hardened silos in mid-continent. All of which could be targeted with Soviet missiles. More importantly, both we and Soviets could target several land-based missile sites and air bases with one missile that had several nuclear warheads.

Before the end of the 'Containment' era we had developed the Triton ballistic missile submarine that is at sea today. The Triton is bigger than its Polaris cousin, has twenty-four nuclear-armed missiles, is quieter, and has a longer-range missile (5000 miles) than those on the Polaris submarines. With only fourteen of these submarines we can maintain a mutual assured destruction (MAD) capability. After the demise of the Soviet Union, Russia and the U.S. agreed that we will no longer target each other. We aim the missiles in the open ocean. We can retarget them anywhere in the world within fifteen minutes and so can the Russians.

What was the impact on the submarine force of winning the cold war? In 1991 the U.S. had ninety-five nuclear attack submarines with the capability to destroy the Soviet submarine force. Today we have fifty-six nuclear attack submarines. During this era of transition, this number will fall further. We are building less than one attack submarine a year because of budget constraints. The Russian submarines are rusting at the piers or sinking. All of a sudden, there were no Russian targets and our submarine force again had to reinvent itself and find a new mission.

In 1945 we had the best submarines in the world to sink surface ships (targets), but there were no targets left. In 1991 we had the best new submarine in the world. It can sink surface ships and other submarines, nuclear or diesel-electric, with devastating ease, and gather intelligence without being detected. It had been designed not only as a submarine killer, but also to be able to land Special Forces, Marines, and Navy SEALs with a little submarine it carries piggyback. The name of that submarine is *Seawolf*. This submarine is quieter at twenty-five knots submerged than any class of nuclear submarines tied up at the pier. As the Soviet Union collapsed, and faced with no opposing superpower at sea, we evolved into the *Virginia* class. We have the submarine and we have a national strategy to use it.

The policy of Preemption has been in effect for just over three years. We are fighting two wars--Afghanistan and Iraq--under this policy. The first, Afghanistan, was fought with air power and cruise missiles launched by submarines, surface ships, and Air Force bombers. Special Forces controlled the warlords' armies from horseback. CIA men with suitcases of money paid the warlords. Marines landed an amphibious force from 700 miles away. The Air Force provided an airborne filling station twenty-four hours a day to the planes from aircraft carriers and the bombers from Diego Garcia. This is the first war that America won without an army. The Army provided the forces to occupy the country after the fighting was over.

The president told the terrorists that there are better ways to meet seventy-one virgins than fighting us. He and Secretary Rumsfeld have made it extremely clear that we would need a different military to fight this kind of war. General Franks wrote the book on how to fight a war in the modern era using every technology available to defeat nations supporting Islamic terrorists. Today our military is writing the book on fighting a terrorist led insurgency in both Afghanistan and Iraq. Our enemies recognize that the only way they can prevail is to use the 'Hate America Firsters' as their fifth column in America. They succeeded in Vietnam. And they think they can do it again. I don't think so. Do you?

The submarine force started getting ready for the preemption wars in both WWII and the Cold War. We were landing Marines in WWII. Today our attack submarines carry cruise missiles that can fly 800 miles and fly in through the window of a building. Four of our Ohio class Trident strategic submarines are currently being converted to carry over 150 cruise missiles each as well as large numbers of special forces and Navy SEALs with their equipment and 'dry' small subs to get them ashore. No other navy in the world has anything like this unique capability.

I learned long time ago to never bet against America. In spite of the efforts of the ACLU and the 'Hate America Firsters', America is still the best hope for the future the world has ever seen. What we have set out to do in Afghanistan and Iraq is vital to the future security of our nation and the rest of the world. We are learning how to find and kill terrorists before they kill us. We have an all-voluntary military that I believe will master this change of policy and learn how to fight a new enemy just as the submariners have done twice in my lifetime. Those of us who served in WWII have been called 'The Greatest Generation'. I believe the current military is the greatest generation. They are all volunteers, men and women, black and white with all shades in between, and they are certainly the brightest and the most patriotic generation in our history.

James Web, Naval Academy class of '68, highly decorated Marine with three tours in Vietnam; Secretary of the Navy in President Reagan's administration; author of bestsellers, said it best in his latest book, *Born Fighting*. His thesis is that the best fighters he served with in Vietnam were 'Red Necks' and 'Georgia Crackers'. He spent thirty years researching where they came from starting with Hadrian's Roman Wall across England. He concluded they were the Scots and Northern Irish who were descendents of the Vikings. These groups arrived in America in the 1700s. They went past the eastern big cities to claim free land in the Appalachian areas and the rural South, and even Texas. They valued their rifles and dared any man or government to try to take them away. These are the largest groups that man the all-volunteer military today. They hated big government and loved fighting. If they continue to have the nation's trust and support, this military will learn how to defeat Islamic terrorism just as we learned how to defeat Germany, Japan, and communism.

Like the veterans who went before them, they will continue to make this nation great and secure. Today I am proud to be an American veteran. I am proud to be a 'Red Neck'. May God bless the American veterans of all wars. May God continue to bless America.

Looking Back at Fast Attack
Jim Frantz, Chief Engineer

Computer games improved significantly in the middle to late nineties. Although the medium was only a decade old, it rapidly went from text only to life-like renditions with sound and cinematic visuals. Two submarine simulations pushed the envelope more than any before, both by publisher Sierra. *Aces of the Deep* dramatized the U-boat Battle of the Atlantic, while *Fast Attack* initiated a new era in modern-era subsims. Developed in 1996 by James R. Jones III and the San Diego-based Software Sorcery development team, *Fast Attack* featured a full complement of stations, including cutting-edge renditions of the broadband and Target Motion Analysis screens. Each function of a *Los Angeles* class submarine was covered in a game that sent the sub skipper to a variety of mission theaters, including the Sea of Japan, Mediterranean, and the Persian Gulf. *Fast Attack* imparted a sense of authenticity unmatched by Microprose's *Red Storm Rising* and EA's *688 Attack Sub*. Jim Frantz, *Fast Attack's* Chief Engineer, "not only oversaw the game design, strategy, scenarios, and controls in the game, but he did it with first-hand knowledge having been a commander of a *Los Angeles* class fast attack sub during his Navy days."

Fast Attack was the creation of James Jones. Software Sorcery had recently completed *Aegis, Guardian of the Fleet* and that game was pretty successful, so another military simulation was a natural successor. There were two other games in development at the time and I, having been a submarine commanding officer until 1985, was the obvious choice to take the lead on *Fast Attack*.

We had several developers working on the various parts of the game. We had Peter Nowak working on the periscope view. Another subsim game had recently received rave reviews for the realistic ocean action and we thought we could at least try to match it. John Knapp was the 'sonar supervisor' and was able to get the sonar's waterfall display to be a pretty darn accurate representation of the real sonar. Ron Yarnall did the 'strip plot' and the fire control and was basically in charge of target motion analysis. Dan Verwys, an ex-Navy P3 crewman, wrote all the scenarios, and Alex 'the Red' Shatsky coded all the enemy behaviors. Ofer Estline modeled the weapons systems. Everyone contributed to the 'voices' used for making reports during the game. I tried to provide as much technical information as possible. Tom Clancy's book *Submarine: A Guided Tour Inside a Nuclear Warship* was used extensively so we could assure James Jones that we weren't using any classified information that would get him in trouble.

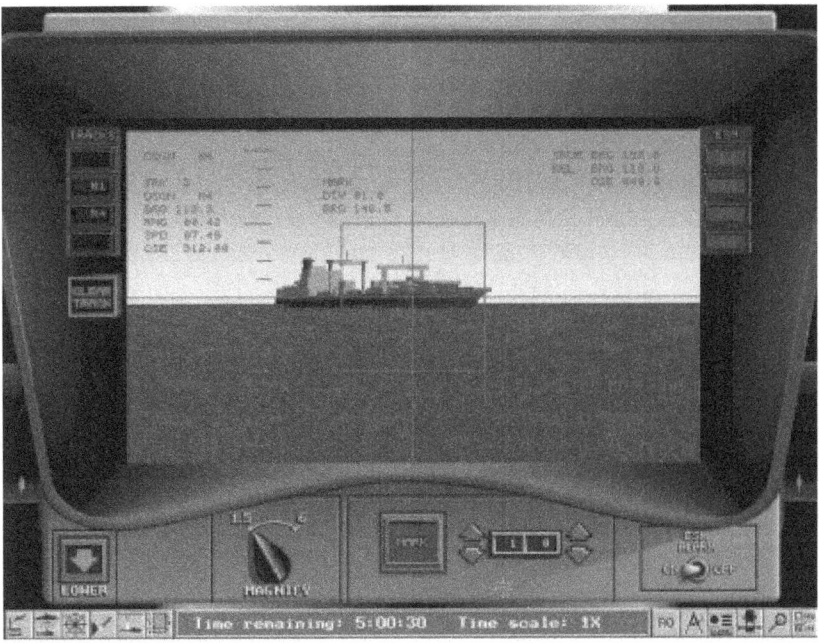

I steered *Fast Attack* toward more of a simulation rather than a game. Early reviews and posts to Sierra's website seemed to appreciate that, but it soon looked like maybe we went the wrong way. There were a few vociferous folks who continually posted 'knocks' about *Fast Attack*—no save game feature, a two-hour mission limit…among other things.

We programmed *Fast Attack* as a DOS game. Window 95 broke toward the end of development while we were finishing up bug fixes and multi-language versions. The timing was a bit unfortunate because we hadn't done much to ensure that the sim worked on a Win95 box, and there were some issues that probably hurt us.

The most difficult aspect of this simulation was teaching the team about how it 'really works'. Hollywood had left a pretty ugly stereotype about submarines. The constant pinging of sonar, the instantaneous travel time for torpedoes, the ease of figuring out the target solution, and instant torpedo reloading were just some of the things we had to 'unlearn'. I wanted the game to make these things more real. We had to introduce bearing errors into sonar that were inversely proportional to signal strength so that fire control and plot would have to make estimates. I didn't want it to be a slam-dunk. I wanted the users to appreciate that it might take hours to make a stealthy approach on a target, especially if there were escorts and limited weapons.

The compromise we decided on was to have three levels of difficulty. That way we wouldn't frustrate the new player who wasn't at all familiar with tactics and whose only experience with submarines was what Hollywood had taught him. At the other end, I knew in my heart that there were going to be some ex-Navy guys, ex-submariners, and even current Navy folks who would be put off by the Hollywood bunk. These people would know that the ocean wasn't transparent, the sky wasn't always clear and that sensors were never perfect. We had to provide realism. We made the sonar bearings on distance contacts vary so the player would have to 'average out' the dots on the fire control or the bearing lines on the strip plot.

The key component of *Fast Attack* was what we called the 'Target Engine'. Here is where we really wished we had the facilities available in today's operating systems. We wanted the Target Engine to run in its own tread, and one of the biggest challenges we faced was to figure out how to simulate this using the facilities of the Watcom tools we were using. At the start of a scenario, all the 'players' got loaded into the TE. This included, own ship, good guys, bad guys, mines, airplanes, etc. Whatever could be detected or run into. When a weapon was launched, it too got loaded into the TE. The TE's function was to report to whomever wanted the information 'who detected whom'. If mines were proximity mines, they would report when they detected another object in the TE. External logic would then decide what to do with that information. In the case of the mine, it would explode and damage would be computed on the object that caused the interaction. Same with torpedoes, missiles, rounds from a gun, etc. The TE was the heart of the whole game.

Dan Verwys would come up with the scenarios. We wanted a mixture of objectives to keep things interesting. So, we had over the horizon (OTH) missions to lay on Tomahawk missiles. We had missions to take on enemy battle groups and missions to sink freighters. There were multiple variants. The weapons load was tailored to the mission. Not all our players were all that happy about what we gave them, but that again was a realism factor. When I was CO, I didn't get a vote on what my ship got loaded with. The type commander specified the ordnance loudout and all the subs of the same class got loaded the same way.

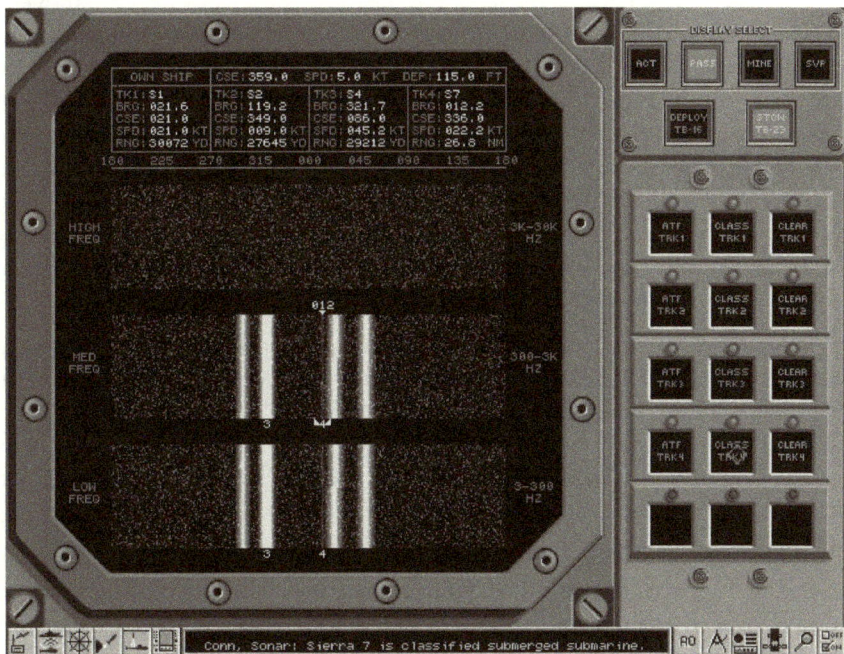

As far as computer assist to the player, the computer did exactly what the computers on the submarine did. The fire control problem is like high school algebra, except you have three unknowns and only two equations. You have to guess (in sub-speak the word is 'constrain') one of the variables. Typically you can guess range. Not accurately, but all you need is 'in torpedo range' or 'not in torpedo range'. Then the computer will help you with the other two variables – course and speed. The default 'guess' for the fire control computer is 10,000 yards.

What we really did to help the novice player in easy mode was to make the torpedoes 'smarter' and give them a wide 'cone of detection' so if the player got close enough, the torpedo would very likely detect and steer correctly without running out of gas. In the advanced difficulty settings the torpedo could be confused by other objects in the TE and the 'cone of detection' was reduced to a more realistic value.

We had planned to add casualties to the mix, things like a sudden fire in the galley, or a hot run in the torpedo room to add excitement and force the player into going to periscope depth or surfacing, but we never had the time to implement these aspects.

I did post a couple of hidden 'features' to Sierra's *Fast Attack* website. These were put in for debugging the TE, but they were never removed. There was a way to get the 'spy satellite' view at any time with a keystroke combination. There was even a way to 'see' all the hidden objects, such as mines. Another 'cheat' allowed you to play a sequence of missions without being penalized for failure to complete primary objectives. This switch allowed testing the medals and promotions aspects of the game without having to plod through every single mission. At the end of a mission, you would still be rebuked for not accomplishing the primary objectives, but as soon as the next mission began, all was forgiven. The game would then proceed as though you had been perfect.

I didn't have many direct dealings with Sierra until just before shipping. At that point, as I recall, money was getting tight, and some developers left to start another company. In the end, I had to fight with Sierra. They wanted more 'Hollywood'–torpedoes were too slow, enemies were too quiet, there weren't enough torpedoes, etc. I basically told them *no*, saying it was too late to make those kinds of changes.

The biggest challenge with Sierra came at the very end. They wanted a German, French and Spanish version of the game. This was very hard and I don't think it ever really happened. We started with sending out all the strings to be translated. The results were almost laughable. The German and French words for most of the things were four times longer that the English words. 'Flood Tube' was *incredibly* long. The remaining artists just threw up their arms when we told them they would have to fit these translated strings into all the artwork!

When Software Sorcery was flush with money, there were all kinds of games on the drawing board. First up was a subsim called *Akula: Red Hunter*. The plan for *Akula* was to be a multiplayer 'partner' to *Fast Attack*. In other words, two players, one with the *Fast Attack* version modified for network play and one with *Akula*, would be able to face off against each other – sub vs. sub. Just before Software Sorcery shut its doors, we had converted the TE to Win95 and made the notifications on detections to go out using IP. It is too bad we were never able to sell the company because multiplayer games (like *Descent*) were in vogue.

I have lost touch with most of the SS team. Ron Yarnall and John Knapp are now my woodworking buddies and we go to woodworking shows a couple times a year. Just recently, Ron literally bumped into *Fast Attack* executive producer Bernie Tyler, otherwise this article would never have been written.

The Krusanov Ultimatum
Andrew Glenn

The man in charge of sixteen nuclear-tipped missiles swayed lazily in his chair, his bloodshot eyes narrowing as he focused on the tumbler of vodka in his hand. With great concentration, Captain First Grade Anatolii Krusanov slowly raised the tumbler to his lips, gulped down its clear contents and slumped back into his chair exhausted.

Watching anxiously from across a small round table, its surface stained and sticky from overfilled glasses was Captain First Grade Mikhail Volodin. The soft-spoken thirty-nine-year-old had seen his friend drunk before, but never like this. Krusanov was a mess; his jowl was carpeted with gray whiskers, his hair untamed and his clothes creased and stained. He looked more like a castaway waiting to be rescued than a respected officer of the Russian Navy.

The claustrophobic, smoke-filled tavern lay on the outskirts of Gadzhiyevo, an isolated port on the Kola Peninsula and home to elements of Russia's Northern Fleet. Its patrons were mostly atomshiks, or nuclear submariners, most of who were drawn by the two full-figured barmaids who serve its overpriced drinks rather than its stifling ambiance.

As Volodin watched Krusanov clumsily poor himself another drink, he recalled how different the circumstances were when they first met at the Dzherzhinsky Naval College in St Petersburg. Volodin was a bright-eyed third-year cadet and Krusanov an ambitious lieutenant commander. Krusanov had been invited to speak about U.S. submarine tactics and Volodin enthusiastically soaked up every word. When Krusanov was given his first command, Volodin had the good fortune to be assigned to his boat as the navigation commander. Despite an age difference of almost ten years, the two men developed a close bond, forged through countless months at sea and numerous close encounters with their Cold War adversaries from NATO.

Since those heady days of the seventies and eighties, much had changed. The Cold War had ended with the collapse of the Soviet Union and the need for a large and expensive navy could no longer be justified. The submarine fleet dwindled from its peak of one hundred and eighty-nine boats to only eighty-six. With funds steadily declining, patrols were reduced by as much as eighty percent. When the submarines put to sea, they often did so without a complete crew. Conscripts and volunteers alike had begun to desert, having not been paid for months, and in many cases were not replaced.

Volodin tried to remain philosophical about the changes as he rose quickly through the ranks. Some believed his good fortune was a gift from his father-in-law, the long-serving Federal Atomic Energy Minister. But those who knew him well quickly dismissed such fallacious observations. Volodin consistently performed above expectations and his dedication and leadership were rewarded with the command of his own nuclear attack submarine when he was only thirty-five.

What did concern him, however, was Krusanov's state of mind. At forty-eight, Krusanov was nearing the end of his career as an atomshik and it was probably just as well. The round-faced, barrel-chested man was slowly being consumed with bitterness as both the woman he loved and the Navy he served wasted away. His wife of twenty-two years was diagnosed with lymphatic cancer and repatriated to a clinic in Saint Petersburg that he could barely afford. Because of the shortage of commanders, the Navy had repeatedly denied him leave. He had not seen her in over six months.

Krusanov had been depressed before, but tonight was different. Tonight he was almost inconsolable. Volodin was shocked when Krusanov eventually told him why. He had just been informed that his wife had passed away. She was only forty-six.

"I just wanted to be there," he mumbled to himself before downing the tumbler of vodka. Krusanov looked up, his eyes reddened by tears and eighty-proof alcohol. "I can't believe they wouldn't let me go. I asked, pleaded, with Fleet HQ to give me just two days..."

"I'm sorry," said Volodin. "It's unforgivable."

"Yes. Unforgivable," repeated Krusanov scornfully.

He poured himself another drink, gulped it down and wiped his mouth with the back of his hand.

"How could this happen, Mikhail? How could those damned fat cats in Moscow let the Navy decay to such a state that they can't let a man visit his dying wife?" He picked up a bottle of vodka by the neck and filled the small tumbler to the brim.

"Have you heard that they are removing the missiles from Konstantin Borisov's boat?"

Volodin found the change in subject curious. But as Krusanov's life revolved as much around the Navy as his wife, it wasn't surprising he wanted to talk about his one remaining passion.

"Yes, I heard," Volodin said sympathetically. "I've not heard why, however."

"So he can carry potatoes and fruit to be sold in some stinking market in Siberia! Potatoes and fruit, Mikhail! Can you believe it? It's... it's a disgrace!"

Krusanov lifted the tumbler to his lips, swallowed its scorching contents and slammed the glass back on the table. His body continued to sway as if he were on the lookout station of his boat. Then, with a groan, he pushed himself out of his chair, wavered for a moment and collapsed back into it with a thud.

"I'm not going to let them get away with this!" he spluttered as he tried to stand again.

Volodin rushed around the table as Krusanov's knees were about to give way a second time. "I think you need to get some sleep, Anatolii," he said as he hooked his arm around Krusanov's broad back to steady him.

Krusanov shook his head. "Not sleep. Must do something to... to stop this madness..."

"I don't think so. About the only thing you'll be doing tonight is snoring. Come on, Anatolii, I'll take you home."

Krusanov leant heavily on Volodin and the two men staggered out of the stuffy tavern and into the frigid October night. The sound of fresh snow crunched underfoot as they made their way to Volodin's four-wheel-drive Lada. He bundled Krusanov into the car and wound his way through the city's quiet streets to the gray apartment complex Krusanov called home. Volodin would have preferred to remain in Gadzhiyevo to make sure Krusanov was all right, but he had orders to sail on the morning tide. As he lowered the barely conscious Krusanov onto his unmade bed, Volodin took some comfort in the knowledge that he was a survivor. No doubt, after the funeral, his friend would take his missile boat back out on patrol and his attention would soon return to the job for which he had been trained: ensuring his submarine remained undetected in the frigid waters of the North Atlantic and, if required, delivering its sixteen SS-N-23 nuclear-tipped missiles on enemy targets.

⊕ ⊕ ⊕

As the sun began to rise over the low, snow-covered hills of the Kola Peninsula, Volodin looked down from his vantage point atop his submarine's sail at the deck below. A handful of black-clad figures carefully maneuvered around the boat's round outer hull untying mooring lines. Sitting low in the water, the seven-thousand-ton vessel had a sleek aerodynamic shape that tapered away to its stern like an elongated teardrop where a distinctive fin protruded high above the water. Christened the *Samara*, the Project 971A boat was the Navy's latest attack submarine. NATO designated it the 'Akula II' as it incorporated several new noise-reduction features that made it much quieter than the first boats in the class. Its low magnetic steel double-hull construction gave it an operational diving depth of over fifteen hundred feet while its one-hundred-and-ninety megawatt nuclear reactor enabled it to slice through the water at a submerged speed of thirty-five knots. Armed with a mix of conventional and nuclear cruise missiles and torpedoes, it was also one of the deadliest attack submarines in the fleet. *Samara* was in excellent condition, having only arrived from the Severodvinsk Shipyard the previous year. Unlike many other boats in the fleet, it would put to sea with a full crew of thirty-one officers and thirty-one enlisted men.

Standing beside Volodin was his second-in-command, Captain Second Grade Boris Mashkov. Although two years younger, Mashkov's lined face and graying beard made him look several years older. On more than one occasion, Mashkov had been mistaken as the boat's captain; an assumption that Volodin often took pleasure in not correcting quickly, much to Mashkov's embarrassment.

The two men were dressed in heavy black overcoats and round lambswool hats. An icy breeze whipped around the submarine, causing the blue and white Russian fleet flag of Saint Andrew behind them to flutter wildly. Mashkov rubbed his gloved hands together and stomped the cold out of his feet on the steel deck.

Volodin looked at his watch and, satisfied that all was ready, gave the

command to caste off all lines.

"*Otdat shvartovy*," repeated Mashkov to the sailors below.

With a splash, the thick ropes were thrown into the oily water and *Samara's* seven-bladed propeller began to churn the water, slowly pushing the boat out of the Murmansk Fjord and towards the open expanse of the Barents Sea.

Following shortly behind *Samara* was a second Akula, *Lynx*. One of the original boats in the class, *Lynx* was commanded by Igor Larionov, a young steely-eyed captain from nearby Arkhangelsk. Larionov had earned a reputation as an aggressive commander whose risk-taking during fleet exercises was often frowned upon by admirals trained to adhere to established doctrine. Yet his crew idolized him. *Lynx* consistently scored the highest number of 'kills' of any submarine in the Northern Fleet while on exercise, earning Larionov the admiration and praise of the boat's company. He rarely socialized with his peers, preferring instead the more raucous company of his younger crew. Larionov's competitive spirit often put him at odds with the likes of Volodin and Krusanov, who were more cautious; a trait he considered to be a sign of weakness. It was therefore with some relief when Larionov opened his orders to discover he would be shadowing a NATO fleet exercise off the coast of Ireland, several hundred miles away from the likes of Volodin and Krusanov. There he and his crew would continue their game of cat and mouse with NATO's submarines without having to share the limelight.

For two uneventful weeks, *Samara* silently patrolled the icy waters of the Barents and Norwegian seas. The only foreign submarine detected was a Norwegian Ula class diesel-electric boat that was steaming southwest, presumably to participate in the NATO exercise. *Samara* followed the Norwegian boat for several hours, breaking contact only to make its regular report to Northern Fleet HQ. Having ascended to a depth of sixty feet, an array of masts and antennae rose above the foamy waves and began to communicate routine information including the boat's location and condition back to Fleet HQ at Severomorsk. The reply that they received, however, was far from routine. As officer of the watch, Mashkov immediately alerted Volodin when he read the deciphered text.

Resting on his bunk, Volodin woke with a start when the sound of Mashkov's voice crackled over the intercom.

"*Captain*, we've received a flash message from Fleet HQ," he said gravely.

Volodin grunted his acknowledgement and threaded his way through the narrow corridor to the Combat Command Center. Located directly beneath the submarine's sail, the crowded pale green compartment was the boat's nerve center. Here, information on what was taking place both inside and outside the submarine was processed, where orders were given to direct its course and depth and to fire its lethal payload. Pipes and wireways hung overhead and the walls were covered with junction boxes, oscilloscopes, dials and stopcocks. Suspended in the center of the compartment was *Samara's* periscope. A grim-looking Mashkov stood beside it and without a word, and handed the red-striped message to his captain. Volodin carefully read the text and looked up at Mashkov.

"Could it be true?" asked Mashkov quietly so none of the other crew could hear the apprehension in his voice.

Volodin wasn't sure what to make of it. "I don't know," he said shaking his head slowly. He re-read the message. It, at least, was unequivocal: *Samara* was ordered to steam immediately at combat speed to within three hundred nautical miles north-east of Iceland where it was to seek out and destroy the missile submarine *Novgorod*.

"That's Anatolii Krusanov's boat, isn't it?" asked Mashkov.

Volodin sighed. "Unfortunately, yes."

"But why would he do such a thing, Mikhail? Why would he threaten to launch his missiles on Moscow?"

"He's gone mad. Anatolii recently lost his wife. Because of crew shortages, he wasn't given permission to visit her in hospital. He blames the government for this and the Navy's decay."

"Do you think he would do it, launch an attack on Moscow?"

Volodin thoughts instantly turned to his family, who had stayed behind in the capital where life was far more comfortable than in Gadzhiyevo. "For the sake of everything we hold dear, Boris, I hope not."

"What are your orders, Mikhail?"

"We must find him. No doubt, most of the fleet will be underway by now with similar orders to sink the *Novgorod*. If we can find Krusanov first, however, we might be able to save him and prevent the unthinkable."

Volodin handed the message back to Mashkov. "Plot a course to this location off Iceland."

"Aye, Captain," said Mashkov as he stepped over to the navigation commander's station. Within moments, a course was plotted. "Recommend course one-nine-two degrees, Captain," said Mashkov. "It should take us no more than six hours to reach the destination at full speed."

Volodin clasped his hands behind the small of his back. "Very well, Mikhail. Steer one-nine-two, depth sixty meters, flank speed."

Samara raced through the depths at thirty-five knots and thanks to favorable currents, arrived at its destination earlier than anticipated. After reducing speed by two-thirds, the submarine began to seek its target. It wasn't long before contact was made.

"Twin screws, bearing zero-four-six degrees!" shouted a young sonar technician as the faint shw-shw-shw of propellers was relayed to him from the bow-mounted sonar. The technician pressed a series of buttons on his console and an onboard computer searched its database of known acoustic signatures. "It's a Project 667 missile boat. High probability it's *Novgorod*."

"Conn, sonar. We've detected a Project 667 missile boat bearing zero-four-six degrees. High probability it's *Novgorod*."

Volodin grunted. Of course it was *Novgorod*. There were no other 667, or Delta IV class boats around for hundreds of miles.

"We're in her baffles, Captain," observed Belyaev. "She doesn't know we're here."

Volodin was well aware that submarines had an audible blind spot where the backwash of their own propeller noise made detecting anything to their rear extremely difficult. To overcome this problem, submariners would clear their baffles by undertaking a series of turns that would enable them to see whether anyone was tailing them. The problem with many Russian commanders, particularly old-school captains like Krusanov, was that they followed a predictable schedule, making it relatively easy for quiet subs like the Akula to follow without detection.

"We better let Krusanov know we've found him," Volodin said casually. He turned to his second-in-command. "Put us on a parallel course, Boris. If his sonar group are on their toes, and I think they would be given the circumstances, then they'll soon know it's us."

"But won't they fear we're here to sink them? If he's threatening Moscow, he's dangerous."

"Perhaps, but we're going to have to risk it. I hope that he'll read our maneuvers correctly and realize we're not trying to blow him out of the water. Besides, I'm counting on Anatolii thinking twice about dispatching his close friend. It's Moscow he has the beef with, not us."

Volodin's strategy was risky. His plan was to surface alongside Krusanov so as not pose a threat and then attempt to find out whether the charges against him were true. If all went well, he may even be able to get Krusanov to return to port where the whole matter could be safely resolved. As electronic communications would be detected, Volodin decided to resort to an old, but much safer, method and use a signal light. The last thing he wanted was Northern Fleet HQ ordering his destruction for not fulfilling his orders and conspiring with a madman.

Mashkov gave his commander an I-hope-you-know-what-you're-doing look and echoed the orders. The crew sprang into action and the submarine began to increase its speed and maneuvered to starboard, placing it on a parallel course to the *Novgorod*. Once in position, Volodin ordered the boat to periscope depth. When the periscope broke through the white-capped waves, he peered into its eyepiece and scanned the gray horizon. Satisfied that they were alone, he ordered the boat to the surface. With a whoosh, compressed air rushed into ballast tanks sandwiched between the two hulls, forcing out tons of seawater in a matter of seconds.

Volodin and Mashkov donned their wet-weather gear and climbed the ladders from the command center to the top of the sail. Even though it was mid-morning, a heavy rain restricted visibility to a few hundred yards. A ten-foot swell crashed over the hull of the submarine, showering Volodin and Mashkov with near-freezing brine.

"*Now to see if we made the correct decision*," shouted Volodin above the din of the brewing storm.

Mashkov nodded warily in agreement and joined his commander in looking to port for signs of the *Novgorod*. After several anxious minutes, Volodin caught the unmistakable wake of a periscope out the corner of his eye. He pointed to the periscope and Mashkov breathed a sigh of relief beneath his damp beard. Moments later the water began to boil and the angular blue-gray sail of the *Novgorod* slowly rose above the waves. It was followed by the long distinctive humpback of the Delta IV class boats that housed its sixteen ballistic missiles. At five hundred and forty feet in length, the missile boat was a hundred feet longer than *Samara* and displaced almost twice as much tonnage. It was an imposing vessel by any standard and its potential to reduce the capital of Russia to rubble made it even more intimidating.

A clutch of six officers emerged from the depths of the boat and took up station on the sail's windswept lookout. Volodin could barely make out their faces through the rain, but knew Krusanov would be there amongst them. Although he felt a great deal of sympathy for Krusanov's cause, Volodin would not hesitate to sink the *Novgorod* if he believed he was about to launch.

Volodin instructed Mashkov to relay a message to the *Novgorod* via the signal light. Choosing his words carefully, he shouted them to Mashkov, who tapped them out on the lantern as fast as his near-frozen fingers allowed.

"ADVISED... YOUR... INTENTION... TO... ATTACK... MOSCOW... IS... THIS... TRUE..."

Moments later, a small light began blinking back at them. Although Volodin could decipher the series of short and long flashes, Mashkov nevertheless translated them aloud.

"DO NOT WISH BUT UNLESS ASSURANCES ARE GIVEN TO ADDRESS THE NAVY'S PLIGHT, WILL."

"Very well," said Volodin over the roar of the storm. "Advise him if I have another way, will he stand down?"

Mashkov flashed the message to the *Novgorod*. After a brief pause, the reply was sent back.

"Da," said Mashkov joyfully. "Krusanov wants to know what you have in mind."

"Tell him I promise to fight this battle with him, but not here. In Moscow, with all the influence I have in the Kremlin to gain a favorable hearing."

"Very good, Captain," yelled Mashkov who then proceeded to tap out the message.

Krusanov's response quickly followed.

"Sir, Krusanov accepts on the condition we escort him back to port. He does not want to risk the lives of his crew any further by making his boat an easy target."

"Fair enough," replied Volodin. "Let the captain know we accept his condition, we will escort Novgorod back to Gadzhiyevo."

As Mashkov began to relay the message, the voice of the sonar group commander pierced the howling wind. "Captain! We have detected a new submerged contact bearing two-six-zero degrees, range seven thousand meters, speed twenty-eight knots."

"*Can you identify it?*" Volodin asked eagerly, hoping that it was a NATO submarine and not a Russian who would almost certainly have the same orders he received six hours earlier.

"One moment, sir," shouted Belyaev.

The minutes it took to confirm the sub's identity felt like an eternity. When Belyaev finally answered, Volodin's heart skipped a beat.

"Captain, it's one of ours; a Project 971 class attack boat."

Aboard the *Lynx*, Larionov looked over the shoulder of a sonar technician and peered at the green zigzagging lines on the oscilloscope. The reading told him the *Novgorod* was traveling at slow speed on the surface. It was a precarious, not to mention curious, situation to be in for a hunted submarine. As far as he was concerned, it just made it all the easier to destroy. What the oscilloscope's lines did not reveal, however, was the presence of *Samara* on the other side of the *Novgorod*. Already one of the quietest submarines in the world, *Samara* was shielded by the larger and noisier Delta IV, making it invisible to *Lynx*.

Larionov snorted his contempt for Krusanov and demanded a firing solution from his combat control commander. Within seconds, a solution had been calculated.

"Now we've got you," he said through gritted teeth. "Lock in the solution, flood tubes two and four and open outer doors."

"Flood tubes two and four and open outer doors, aye, sir," echoed the combat control commander, who relayed the order to the torpedo group in the forward compartment. Barely a minute later, two red lights illuminated on the fire control panel indicating the two twenty-one inch Set-65 torpedoes were ready to fire.

Volodin was alerting Krusanov to the fact that they had company when Belyaev shouted over the intercom that *Lynx* had just opened its outer torpedo doors. Volodin instinctively gave the order to dive. Opening outer torpedo doors meant only one thing in this situation and the worst place to be was on the surface traveling at slow speed. Mashkov slid down the ladder into the sail. Volodin followed and secured the hatch behind him. The command center was filled with a cacophony of alarms and men shouting orders. Stern-faced officers and wide-eyed sailors busied themselves about their stations. Volodin and Mashkov took up their positions near the periscope and ordered the boat to three hundred meters. Volodin realized he needed to get a message off to *Lynx* warning it not to shoot. Yet no sooner had he reached for the handset to contact the communication commander when a sonar technician heard the distinctive high-pitched whine of a torpedo's counter-rotating screws through his headset.

"Torpedo, torpedo, torpedo!" shouted the lanky technician, sending a chill down the spine of everyone within earshot. "I have one, no, two incoming torpedoes, bearing two-seven-zero degrees, range five thousand, five hundred meters, speed forty knots!"

"Right full rudder, flank speed, depth one hundred meters," said

Volodin in a calm, reassuring voice. The orders were echoed instinctively by the crew as they slowly coaxed the submarine into what they hoped was a less vulnerable position. "Prepare countermeasures."

Samara was fitted with several countermeasure features to detect and defeat electronic and torpedo attacks. One of those was a torpedo-sized noise simulation decoy designed to lure an approaching torpedo away from a submarine by making it appear as though it was a more attractive target. The newly installed decoy was about to get its first operational test. Volodin prayed it lived up to its designers' promises.

"Torpedoes now bearing three-zero-five degrees, range four thousand two hundred meters."

The tension inside the command center was palpable. As the submarine turned and dived to put as much distance as possible between it and the torpedoes, Volodin's knuckles turned white as he hung on to the chart table behind him. Mashkov gripped a stopwatch in one hand and steadied himself against a bulkhead with his free hand.

"Three minutes to impact," he said grimly.

Volodin ordered the countermeasures fired. A jet of compressed air shot the decoys out of the submarine with a loud whoosh. They activated immediately, mimicking the acoustic signature of the substantially larger Typhoon class missile submarine.

Volodin ordered a course change so they would be nowhere near the decoys when the torpedoes arrived. *Samara* continued to dive and the torpedoes followed it down into the crushing depths, their onboard sonars now actively pinging the darkness in search of a target. As the first torpedo drew closer, it detected the decoy and began to alter course. A sonar technician aboard *Samara* jubilantly announced the torpedo had been duped and was now speeding off in the opposite direction. A few seconds later the second torpedo approached the decoys. For a moment, it appeared as though it too would veer off course, but as *Samara* turned at thirty-five knots, its propeller began to cavitate. The sudden formation and collapse of low-pressure bubbles caused by the boat's fast-spinning propeller was sufficiently loud to attract the torpedo's attention away from the noise of the decoy. The twin planes at the rear of the Set-65 torpedo controlling direction and depth suddenly changed, swinging the deadly weapon around until it was on a collision course with *Samara*.

"Two minutes to impact," said Mashkov, his eyes transfixed on the stopwatch as it continued its deadly countdown. Beads of perspiration formed on his lined forehead and slowly slid down his temples. He nervously wiped them away with the palm of his hand.

"Sound collision alarm," ordered Volodin solemnly.

A loud klaxon reverberated throughout the submarine and crewmembers sealed compartment doors and braced themselves for the inevitable.

"*One minute to impact.*"

The men in the command center looked anxiously at each other as the high-pitched sound of the torpedo could be heard approaching to port. Chests were crossed and lips silently asked for deliverance. Only a miracle would save them from the destructive force of the torpedo's three-hundred-and-fifty-two-pound warhead.

With only seconds until impact, it looked as though their prayers would be answered. Belyaev reported the sound of a second submerged contact closing fast — the *Novgorod*. It was steaming at full speed and heading almost directly for *Samara*. At first, Volodin thought Krusanov was positioning his boat in the path of the oncoming torpedo to draw it away. But with such a short distance between the submarines and the torpedo, he soon realized the maneuver wouldn't work. Then it dawned on him. Krusanov wasn't hoping to lure the torpedo away. He was placing himself in front of *Samara* as a shield.

Before Volodin could react, there was a gut-wrenching explosion as the torpedo struck the *Novgorod's* bow. For a split second, a brilliant flash illuminated the depths as the warhead disintegrated the boat's forward compartments. The shockwave from the explosion vigorously shook the *Novgorod* from bow to stern, causing further catastrophic damage along the length of the boat. The shockwave soon hit the rear of *Samara*, flinging its crew and everything else that wasn't bolted or tied down from one side of the boat to the other. Valves already under immense pressure burst, forcing gallons of water into the rear compartment. Junction boxes shorted and the emergency lighting self-activated, bathing the smoke-filled compartment in an eerie amber light. Crewmembers staggered to their feet and rushed to seal the ruptured pipes and extinguish electrical fires.

"Damage report," demanded Volodin as he wiped blood from a gash on his forehead after being thrown against the edge of a control station.

Mashkov picked up a large green telephone and requested a damage report. A breathless officer responded, updating Mashkov on the crew's efforts to contain the damage.

"The reactor is stable and still online. Flooding and electrical fire in the rear compartment," replied Mashkov. "Both are under control. Some minor crew injuries, but no other significant damage."

"Very good. Take the boat to communications depth. We need to make sure Larionov doesn't try to sink us again and that Fleet HQ is aware the *Novgorod* has been hit."

As *Samara* slowly ascended to sixty feet below the waves, the sound of the stricken *Novgorod* could be heard succumbing to the extreme pressure of the deep water. Sonar technicians detected the faint popping noises of door seals failing and the louder crashes of bulkheads beginning to collapse and seawater rushed into occupy the space. The screeching of steel twisting under fierce stress gradually faded as the submarine corkscrewed to the bottom. Moments later, the only sound that could be detected above the background noise of the restless ocean was the rhythmic swooshing sound of propellers as the two Akulas traveled towards the surface.

⊕   ⊕   ⊕

At the windswept and snow-covered headquarters of the Northern Fleet in Severomorsk, Volodin stood patiently outside the fleet commander's office. He looked out at the white parade ground below and watched the last of the mourners leave the ceremony that was held in honor of the *Novgorod's* one hundred and twenty crewmembers. At the far end of the parade ground stood the towering statue of a Second World War Soviet marine whose heroic poise was immortalized in mid-step as if the giant was snap-frozen by the frigid arctic winds. Volodin wondered how many ceremonies dedicated to the loss of Russian submariners it had witnessed. Far too many he concluded.

"Admiral Chernov is ready to see you now, Captain," said a young, rosy-cheeked lieutenant.

Volodin was shown into a large wood-paneled office, lavishly decorated with antique furniture and heavy crimson drapes. Admiral Viktor Chernov sat stiffly behind an oak desk and sipped tea from a white bone china cup. He waved Volodin in and motioned to a single chair in front of the desk with a nod of his head.

"A tragic business," said Chernov. "Very tragic."

"Yes, sir," agreed Volodin as he pulled up the chair.

Chernov took another sip of his tea, placed the cup onto its saucer and pushed it to one side. "Mikhail," he began in a somber tone, "I've asked

you here to discuss in private a number of issues concerning this terrible loss."

"Of course, sir."

"First, I want to say that I share the views you so forthrightly expressed about the Navy's malaise at last week's hearing into the *Novgorod* incident. I don't think I would have put them quite so bluntly myself, however," he said.

Chernov reached across the desk, opened a small silver box and plucked out a cigarette that he slid into a short holder and lit with a heavy crystal lighter.

"Anyway," he said as a twisting column of blue smoke ascended to the high, ornately decorated ceiling. "I also wanted to reaffirm my confidence in you, despite the hearing's conclusion that you should never have taken the risks you did. Indeed, some of the panel wanted to see charges laid against you for disobeying orders. Apparently, however, they were encouraged to revise their decision by a higher authority." Chernov reached across the table and flicked ash into a crystal ashtray. "It would appear that you have a well-connected guardian angel, Mikhail."

Volodin smiled solemnly.

"Speaking of which, you may be interested to learn that a Cabinet reshuffle is in the offing. My sources tell me that the influential Atomic Energy portfolio is to be given to a relatively junior Cabinet member. It would appear that your guardian angel's wings are about to be clipped."

Volodin was shocked. It was the first he had heard of it.

"What's more, there's likely to be another reduction in the defense budget. Apparently, the president wants to progress his health reform agenda and sees defense as soft target. Rather than convincing him that more money needs to be spent on the Navy, the *Novgorod* incident has somehow convinced him that we are just a bunch of high-strung boys with expensive, not to mention deadly, toys who need to be reigned in. Which makes what I have to say next all the more difficult."

Volodin watched Chernov take a final puff from his cigarette before stabbing it out in the ashtray.

"You were planning to take leave to visit your family in Moscow next week, weren't you?"

"That's right. I haven't seen them now for eight months."

"I'm sorry to say that all leave has been cancelled. Despite the fact that

Moscow has little confidence in what we do, they still expect us to patrol the ocean. As I'm no longer getting replacement crews, I have no choice but to ask you to put to sea again."

Chernov stood to his feet, signaling that the meeting was over. "I'm sure we'll have this matter sorted out by the time you come back," he said casually.

"I'm sure you will, sir," replied Volodin flatly. He put on his cap, saluted and turned to exit the office. As he reached for the door, Chernov took a step forward and put his hand on Volodin's shoulder.

"Mikhail, now, you won't do anything precipitant, will you?" he asked through a forced smile. "We can't afford another... embarrassing incident."

Volodin stared at Chernov's hand until he removed it.

"The thought never occurred to me."

Puppies of the Pacific
Chris Weisensel

Lt. Commander Fred Johnson saluted COMSUBPAC as he boarded his new boat, the *USS Goldfish*. Fresh from new construction, he was eager to see what the best of naval design was capable of. The protocols of leaving port were completed, and he gave the order to get underway. The boat was kept at slow speed until the channel was cleared. He then went below to enter the departure on March 6th, 1943 in his log and to check the navigator's course to Midway.

Johnson looked aft of his boat at Pearl Harbor as it shrank into the horizon. It struck him that they were all very well cared for while they were there. He would miss the luaus, the food, and the chats with other sub skippers about what worked and what did not. Now, it was just him, his boat, and his crew. And an ocean full of Japs. With that he told himself to turn around and look forward instead of back. As he moved to the bridge, a seagull soiled the spot he had just stood. For a split second he considered manning the antiaircraft guns, but then thought better of it. He checked the bird for meatballs, and then chalked it up to good luck. He would take every bit of luck he could get.

His executive officer, Lt. Dan Parker, saw the near miss and turned to his skipper with a smirk on his face. "Close call, sir?"

Johnson shook his head. His XO seemed to see everything. It was a talent he was glad to have on board, but he was also sure the entire crew would hear about a seagull almost landing a direct hit on him. "Let's just hope we're this elusive when we're out in Empire waters."

"Aye, sir," came Parker's automatic response. He was pleased the boss could take a joke. He had served under too many self-important jerks to count, and he knew how it hurt morale.

"Skipper, we've got a contact on SD radar," reported the officer of the deck, Lt. J. G. Robert Jamison. Johnson walked into the control room ordering a crash dive. He knew it was probably a PBY search plane from Midway, but it made good practice anyway. He watched as the crew went about their routine of diving the boat. They seemed faster this time.

"Sir?" Parker said to get Johnson's attention.

"How'd they do, Dan?" asked Johnson.

"They shaved five seconds off their best, diving the boat in forty-three seconds," reported Parker with just a hint of pride in his voice.

"That's good news. We should be to Midway in about twenty-eight hours," replied Johnson. The quick dive boosted his confidence in his crew, knowing that they might be able to bail him out of trouble if he made a mistake.

First it was Pearl, now it was Midway that disappeared from the horizon. What lay in front of them now was the vast expanse of the Pacific. After two days of sailing towards their destination, Johnson decided it was time to let the crew in on their patrol area. Once he announced that they were on their way to the hunting grounds between Luzon and Formosa, the crew seemed to perk up. At least it was a worthwhile mission. It would also, however, take a while to get there. As routine started to establish itself for the journey, Johnson made it his purpose to break up the routine with drills as best he could. They had to be ready for anything.

A week out of Midway, the weather started to deteriorate. Johnson knew that the reduced speed he would get from his boat during the storm would shorten his time on station, but he was glad that he could dive if it got unbearable. He made a quick mention to keep an eye on the exhaust vents for the diesel engines so water did not get in. He then went up to the bridge to take a look around. The rain was stinging sharp. The temperature was in the mid-fifties, and the clouds were low and dark. He looked at his watch to confirm it was 1300 local time, but had the red lights turned on due to the low light condition. He asked the lookouts how they were doing. They were wet, but alert. He had to do a double take because, for a moment, he could have sworn he saw a railing through the port lookout.

Chances of an encounter with the enemy here were remote, but not impossible. There was one thing he did not understand though. He could hear thunder, but he could not see any lightning. No matter how loud the thunder, or what direction it was coming from, the lightning that caused it was invisible to him.

The storm abated after a couple of days, and they were back up to speed. It was a little over two weeks from Midway to their patrol region in good weather, but the storm slowed them down. The first week was, not surprisingly, uneventful. As they neared their destination, the very real possibility of air attack heightened their senses. The SD radar existed to notify the boat of an incoming aircraft. It was a good system, but all the sub commanders that he had talked to in Pearl had decided to submerge their boats during daylight hours when in range of Japanese air cover. It was a prudent measure, and he intended to follow suit as soon as he was within one thousand nautical miles of Luzon. It might be boring at first, but sinking every ship he saw was a secondary mission behind getting his crew and boat home.

The noise of diving woke Johnson out of a light sleep. A quick glance and he was able to note that the watch was taking her down at 0200 local time without consulting him. Jamison was on his way to report to Johnson, who popped out of his cabin and met him halfway.

"Sir," he whispered, "we've got a large freighter off the port bow at a range of about two thousand yards, she's right on top of us. Visibility is down, I ordered the boat to periscope depth as soon as the lookouts called her out." It was obvious by his expression that he was hoping that the skipper would approve of his decision.

"Good call, Bob. Sound general quarters and get the attack team up here right away. I'm going up to take a look for myself. Just one thing, there was no radar contact?" Johnson queried his junior officer.

"No, sir," replied Jamison. "No idea why either."

"Well, we can figure that out later."

Johnson climbed the ladder to the conning tower to get a handle on the situation. Own ship course was 240. Own ship speed was three knots. Current depth was sixty feet. Charts said they were in deep water.

"Up scope," barked Johnson. He waited as the periscope traveled up above the surface. He squinted as the scope broke the surface. There it was. He estimated it was a freighter in the five-thousand-ton range. He took measurements and relayed them to the now assembled attack team. "Target range, one-seven-five-oh…angle on bow, 078. Down scope."

He had the torpedo crew set the weapons to a depth of twenty feet, using the contact detonator. When the settings were made, he ordered the outer doors opened on tubes one and two. It was then time to update the track of the target.

"Up scope." Now that he was used to the darkness, he quickly picked up the target. "Target range one-five-double-oh…angle on bow 084…down scope."

"Estimated speed is between five and seven knots, sir," reported Parker, a little uncertainly.

"Very well, prepare to fire tubes one and two. Set a spread of two degrees," ordered Johnson.

"Torpedoes ready, sir," was the quick response.

"Up scope," Johnson said again. It should be the last time before impact, he told himself. He wanted to keep the scope up the whole time to watch, but he fought the urge.

One last set of measurements confirmed the attack team's track, and the scope went down again. At a range of twelve hundred yards, the *USS Goldfish* fired her first war shots. The boat shuddered slightly as the torpedoes were ejected from their tubes with compressed air. At forty-six knots, they closed on their target quickly.

The entire will of the crew seemed to carry the torpedoes to the target, and the seconds seemed to slow to a crawl.

"Torpedo impact!" reported Parker, although the loud report reverberating through the hull meant that the announcement was not really necessary. Four seconds later, a second explosion signaled the second torpedo had also found its mark.

"Up scope," Johnson said with a smile visible to all. He found the target with fire on its starboard side, but no seeming loss of buoyancy. *That's odd*, he thought. He looked along her waterline for signs of damage. Sure enough, he found scoring consistent with two separate impacts. The freighter continued to steam on, however. He should be more patient, he told himself, but there is no way he was going to let this ship get away. He set up for another two torpedoes. He wanted to put one right into the freighter's screws this time, so she could no longer run. With the evasive maneuvers the freighter was now making, he was sure one of the two torpedoes he was about to fire would miss. It bothered him a little, because he did not want to waste his boat's firepower, but that firepower existed to sink ships. Leaving a ship damaged, no matter how severe, left open the possibility of repair, and, more importantly, prevented the tonnage from being counted in their patrol log.

With the second salvo ready, Johnson ordered the firing of the second salvo. The first torpedo missed to starboard, but the second one slammed home into the ships screws and rudders. *That ought to stop it*, he thought. Sure enough, the ship slowed and was eventually adrift. Through the periscope he studied the ship, looking to see if it was fitted with any armaments. He found a gun on the bow and the stern. Probably in the four or five-inch range, he surmised. He had no idea how well trained the crew was at using the guns, so he decided to stay submerged for now. He kept the ship under observation and noted that it was starting to get heavy in the stern.

Another hour passed, and the ship showed no change. Johnson was a bit annoyed. He knew that it had probably sent out a radio call, and it would only be so long before escorts would come. He decided to amplify the ship's problems by hitting it on the same side and, hopefully, one compartment forward of what he figured was a flooded engine room. He would fire only one torpedo this time. Again *Goldfish* shook slightly as the weapon left on its collision course. He could indulge himself this time, and he watched the torpedo track straight to the freighter. A plume of water erupted right where he aimed, and the sound was almost immediate. A quick check in his head said he had fired five of his twenty-four torpedoes for four hits. That was entirely too many for a five-thousand-ton ship. He wondered if it was empty, or what it could possibly be carrying.

Finally, ten minutes later, the freighter's stern was awash. With the bow elevated and the stern mount in the water, it was time to finish the vessel off with gunfire. "Surface the boat! Battle stations, gun!" he ordered. The crew leapt to comply and the vessel broached the surface smartly. The hatch popped open and the gunners took their position. "Keep hitting the starboard side, fire when ready," he barely finished when the first round was on its way to the target. The range was less than a thousand yards, and neither vessel was moving under power. It was an easy shot.

Rounds began to slam home. Most struck near the waterline, but the elevation of a few were off due to wave action. After twenty rounds the freighter started to list to starboard in a meaningful way. Johnson ordered a cease-fire to save ammunition. The list aboard the freighter continued to exacerbate and, finally, the enemy ship slipped beneath the waves. A lifeboat from the freighter was in the water and it looked to him like there were about a dozen men on board, but neither the Americans nor the

Japanese seemed at all interested in having *Goldfish* pick up the crew of the now destroyed freighter.

It was then that a crate from the merchant ship floated up to *Goldfish*. The gun crew hauled it on deck. It was opened and the contents caused uproarious laughter amongst those on deck. Johnson grabbed one and threw the rest of the crate overboard. He ordered ahead full to clear the datum of the sinking and then went below. "Check this out, men," still laughing. "We just spent five torpedoes and a bunch of deck gun ammo to sink a ship full of lifejackets!"

The crew in the control room started laughing, too. Already they had a good story and a trophy. Johnson motioned Jamison over. "Bob, go tell the guys in the forward torpedo room that they can find out what's so funny as soon as all tubes are reloaded."

"Aye, sir," Jamison said, still chuckling.

"Bob, one more thing."

"Yes, Skipper?"

"When that's done let's get to work on that radar."

"Right away, sir"

⊕ ⊕ ⊕

After two days of work they discovered that, while it looked like it was fully operational, the radar would only detect contacts when the submarine's course was northerly. If *Goldfish* was on a southerly course, the radar would detect nothing. Johnson had a radio message with the discovery sent to command. The response came back a little sooner than expected.

"Message from command, sir" the radio operator said as he handed the sheet to Johnson.

"Thanks, Pat," Johnson answered, eager to see the reply.

PROBLEM NOTED ON OTHER BOATS X SOLUTION FOUND TO MODIFY EXISTING SETS X MUST BE IN PORT TO ACTIVATE MOD X COMSUBPAC

It was a somewhat cryptic response. Johnson was somewhat relieved to find that they were not the only sub with the problem. He was also, however, a little irritated that the system was not fully tested before it was released for production. Irritation or not, they were still in the Pacific, and had to learn to use what they had to its best advantage. Submarines went to war without radar for over a year, so half a radar was better than none.

Three days of empty sea followed the sinking of their first ship. The fourth day seemed to be more of the same, but after the sunset, a radar contact was made. Johnson was there to begin prosecution of the contact, which was almost bow on at a range of twenty-five thousand yards. *Goldfish* was headed almost due north, and Johnson hoped that the contact was moving south. The big spot on the PPI scope faded after the energy of the radar return dissipated, revealing that it was actually five smaller spots. They were lined up vertically, with the southernmost contact farther ahead of the other four. Johnson immediately extrapolated that it was a convoy of four merchants led by a single escort. They were a little more than twelve nautical miles away, and on reciprocal courses. *Goldfish* was already in position, and things were going to happen very quickly.

Johnson quickly called general quarters and had the boat submerge to radar depth so he could keep track of them in case they were to zig or zag. He wanted to hit the escort first with the element of surprise, and then the rest of the convoy would be at his disposal. He used the radar sparingly, with single sweeps every five minutes. They did not break their formation, so he was assuming they were not detecting his emissions.

A bright, full moon illuminated the enemy ships at three miles out, and he kept the radar off for good. *Goldfish* slid down to periscope depth, and Johnson, while noticing the merchant vessels, focused on the escort. He needed to identify it so that he could get an accurate range with the stadiameter. None of the destroyers in the book looked quite right. He started to look at some of the other auxiliary vessels and found one that matched. It was a minesweeper. At five hundred tons she was hardly worth a torpedo, but she was also the only thing between him and the convoy. He could avoid the minesweeper and attack the merchant ships, but he doubted he could sink them all in a single salvo, and the minesweeper had enough depth charges to keep him busy while the other merchant ships made their escape. Worthwhile or not, the minesweeper was getting a torpedo.

He started taking range measurements, and the attack team started a track. The convoy was headed south-southwest at roughly six knots. He took a quick look at the merchants and found that one was a bit bigger than the rest, but none of them was very impressive. Still, it was tonnage.

The range on the minesweeper shrank quickly, and Johnson maneuvered to take a stern shot at her from a range of one thousand yards. He aimed at the bow because the instinct would be to speed up if the torpedo were spotted. He was torn between firing two torpedoes to be sure and the waste of another precious torpedo. He decided the second torpedo would be worth it to protect his crew from a miscalculation on his part. Tubes seven and eight were made ready and, when the time was right, they were fired.

Johnson waited the seconds out, hoping against hope that the minesweeper was not very observant. The speed of the torpedo was an advantage, but the steam wake, especially at night, was quite visible. In this case, the torpedo speed won out, and the report of a torpedo impact thundered through the water. The scope went back up and Johnson saw the minesweeper with its stern blown off. The second torpedo sped into the night.

Expecting the convoy to scatter, Johnson was amazed when he looked through the scope and found them slowing down. He could not believe his luck. Maybe they thought they had stumbled into a minefield. He decided to save the conjecture for a later date and acted to take full advantage of the situation in front of him. He increased speed and maneuvered *Goldfish* so that he had the starboard broadsides of all four ships lined up in front of him. He took aim at the lead ship and loosed a torpedo. A minute later the ship broke in half on impact. The sequence repeated on the second ship in line, although a slight miscalculation meant that the torpedo hit farther forward than Johnson would have liked. The entire bow was blown off, however, and it sank rapidly, beating the first merchant ship to the bottom. The sea was dotted with lifeboats.

The third ship in line was the largest of the three, but was no larger than the single ship they had sunk four days earlier. This one, however, was not carrying lifejackets. As a torpedo hit just aft of the mainmast the ship erupted in flames. Secondary explosions shook the stricken ship, which started to sink by the bow with a list to starboard. The final ship made no move to avoid its fate, and it too broke in half.

Sinking one ship after another was becoming methodical. Johnson was baffled by the lack of action on the part of the merchant ships, but it was not his job to understand their behavior. That was for people higher up the chain of command. His job was to sink ships and write reports.

Thirteen torpedoes left, Johnson thought to himself. *Goldfish* had fired eleven and sunk six ships for roughly fifteen thousand-tons. That was a good patrol, but he wanted more.

Another week went by, and the lack of diesel fuel would soon make his decision to leave for him. It hardly seemed like enough time on patrol, but getting a tow back to port was not a good way to impress the superiors.

After two more days of nothing, Johnson finally decided to bring his boat back. He considered a quick refit at Midway, but getting the radar fixed required docking at Pearl. With the results of his patrol, it was possible he had enough renown for an upgrade as well. So he chose to return to Pearl, stopping at Midway first for fuel and fresh provisions. One day into the return trip they received a radio message that would delay their departure.

TO USS GOLDFISH X

ENEMY COMMUNICATIONS INDICATE SORTIE OF CARRIER GROUP FROM MAINLAND JAPAN TO BASE AT TRUK X INTERCEPT AND DESTROY CARRIERS X BE MORE AGGRESSIVE X COMSUBPAC

Maybe they had not received his last battle report, he thought, somewhat irritated. After consulting with his navigator, they plotted a slow course that maximized fuel efficiency while still keeping *Goldfish* on a northerly course as much as possible. He mused that some of his colleagues complained about getting the same orders over and over. Even if the orders are the same, each patrol is different.

The entire crew was on edge. Another two days of nothing. Waiting for an opportunity this big was exciting, but the uncertainty of whether they could achieve a firing position, as well as the fuel situation, was causing some anxiety. Then it happened.

"Contact on SD radar, bearing three-four-seven," reported the radar operator.

"Periscope depth," ordered Johnson, waiting for the chance of a lifetime. It was 1400 local time, and *Goldfish* was at radar depth. Johnson raised the scope and looked down the indicated bearing. The type of aircraft was of immense interest to him. If it was a flying boat, nothing changed. If it was a carrier-based plane, or cruiser-based scout, then they were close. It took a few minutes for the plane to close the gap between the radar detection range and the range at which it was visible to the eye.

It was a cruiser-based scout plane. Maybe it was patrolling, and maybe it was training. Either way the scope went down as soon as Johnson made the identification. If they were detected, any chance they had at an intercept would evaporate. The tension increased, as everyone now knew there was big game afoot.

Four hours later, a call came from sonar. "Sonar contact. Warship bearing two-nine-three." Johnson thought about going up to radar depth to get a picture of the task force, as well as a range, but decided against it. If any gear in the Japanese fleet were capable of detecting radar, it would be installed in a carrier task force. As he waited to develop the contact further, more contacts were audible on the sonar. Johnson called for general quarters and went to the hydrophones to listen for himself. It sounded like a thundering herd. The different screws sounded different, but he could not tell the difference between a cruiser and a carrier.

Johnson raised the periscope and looked for the ships. They would be easier to spot with the sun low in the horizon, providing a silhouette. The scope went back down. This was repeated two more times at intervals of five minutes. The bearings generated by sonar were moving left to right ever so slowly. The slow bearing rate gave Johnson hope for the intercept. The scope went up again. There, on the horizon, he saw half a dozen ships. One had a flat top. He immediately focused on it and started to give bearings to the assembled attack team.

Ten minutes later, he had a good handle on the situation. There were a total of twelve ships he could see. Two were carriers, one was a battleship, four were cruisers, and five were destroyers. Of the two carriers, one looked like a light carrier, while the other looked like a fleet carrier. Their current course appeared to be 100 degrees, and they were making roughly twenty knots. Johnson looked at the plot. It put them at a range of two thousand yards for the salvo. He considered speeding up to close on the track, but decided it would be better to keep some flexibility if they changed course. Their current course was too far north for Truk. He expected them to turn south at some point.

Johnson was worried he would wear out the periscope, but he kept giving bearings and ranges to the attack team. The enemy carrier was making nineteen knots and their range was now seven thousand yards. Then they changed course. Johnson had to wait until the course change was completed. Then the scope went back up. Their new course was 150 degrees. Johnson changed *Goldfish* to a westerly course, and all the computation started again. Target interception was now at two thousand five hundred yards. Still farther than Johnson would have liked, but not an impossible shot either. He had the torpedoes set to twenty feet and contact exploders. The outer doors on all ten tubes were opened. He was going to shoot every torpedo he had loaded at the carrier. It would leave him with three torpedoes, but he was going home after this anyway. With the position keeper engaged, he observed the target to confirm the firing solution.

With the first destroyers within three thousand yards, Johnson ordered silent running. The crew could hear their hearts beating. The clock ticked slowly. Then it was time.

"Fire tubes one through six," whispered Johnson. The shuddering of *Goldfish* was now becoming welcome. Once the final bow torpedo left its nest, the submarine went to left full rudder to unmask her stern tubes. "Fire tubes seven through ten," Johnson ordered again. Four more torpedoes burst from *Goldfish* towards the enemy. "Take her deep. New depth five-zero-zero," Johnson ordered, taking the submarine to evasion mode, heading away from the task force and into the depths.

Everyone was waiting during the crucial seconds between firing and impact. Seventy-three seconds. Fifty-two seconds. With thirty-one seconds left on the count, they started to hear pinging. Did the carrier see the torpedoes in time? Big ships hardly have the best turning radius. Time ran to zero. No explosions yet. Time plus three. Time plus five. Suddenly an explosion. There was euphoria from the crew. Two more hits after that. Secondary explosions were heard on sonar. He would have given anything to see what was going on up there. Then one of the four stern torpedoes found a target. Who got hit by what was a mystery as they tried to escape the inevitable counterattack.

Two depth charges exploded behind them. Johnson ordered three degrees right rudder to keep his course changing. Three more explosions echoed through the hull. They were closer. More pinging was heard, but it sounded different. *Goldfish* had passed through a thermal layer at three hundred and fifty feet. It was reflecting some of the sonar energy from the Japanese destroyers. They could hear pinging from two separate destroyers. At least one more was probably listening for noise from the submarine.

"Depth charges in the water!"

If he could hear the splashes, they were close enough to be a concern. "Take her fifty feet deeper."

Johnson was standing over the helm, willing the submarine through the ocean. Four depth charges hammered at the hull of *Goldfish*. The crew was startled, but no damage was taken. Four more charges were a little farther away off the port side. The crew could feel the submarine shrink under the pressure of the sea, knowing that any leak at this depth could be fatal. The proximity of the depth charging fell away, however, and the tension in the submarine began to subside. They stayed down for another three hours before coming up to periscope depth. Sonar had heard destroyers departing, but Johnson had to know if there were any drifting quietly, waiting to pounce. A sweep with the periscope found no destroyers, but there was one burning carrier. He ordered *Goldfish* up to radar depth. After ensuring that he was on a northerly heading, he ordered a single sweep. The only contact was the carrier. The set was turned off and the carrier was observed. It was listing heavily to port. Another half hour and she slipped beneath the waves.

USS Goldfish rose to the surface and began to recharge her overworked batteries. Johnson set course for Midway and told the engine gang to keep the boat at ten knots or they might not make it home. He reflected on how easy the evasion was and how quickly the Japanese escorts had given up. He was glad he was not fighting the British in the Atlantic. They probably would have kept him down until the batteries went flat. There would be time to reflect on that later. It was time for him to sleep.

⊕ ⊕ ⊕

The submarine's return voyage was long and uneventful. They docked at Midway with five percent fuel. After taking on fuel and some provisions, they sailed for Pearl. Less than a week later they were turning into the channel. The men were exhausted, but extremely confident. Best of all, Johnson thought, they were alive. He was satisfied that he had taken no unnecessary risks, and had a very successful patrol. He was curious if he would get retired after his first patrol in *Goldfish*. He had heard of some skippers being retired after their first patrol in a new boat. That made little sense to him, but neither did half-working radars, storms without lightning, convoys stopping while under attack, destroyers that gave up too easily, and the crewmen that he could swear were occasionally transparent.

Next time will be harder, he thought. It always gets harder. He got back to base and started hearing rumors of the *Trigger Maru*. Its success would make life more difficult for the rest. Their experience would serve them well, but they would have to grow to survive.

Do You Believe in Miracles, Jake?

Alan Bradbury

The bow of *U-534* towered high above Jake as he wandered around her, taking in the smooth whale-like beauty of the old U-boat's bow planes in the early-morning winter sunlight. Even in her dilapidated state, the boat still exuded power and grace.

"She's a beautiful thing, isn't she?" Jake opined, as a stranger on the other side of the boat emerged, similarly entranced.

"I'm sorry?" the stranger replied as he turned around, revealing that his black suit was the outfit of a priest.

"I was saying she's a beautiful thing, Padre, although judging by your profession—and her purpose—I guess maybe you don't entirely agree."

"Oh, that's all right, son, even a thing of war can still be a thing of beauty, it's only what it can do that might be ugly," replied the priest.

"I guess that's true enough, even if it isn't fighting the good fight," Jake replied. "Name's Jake Finley."

The priest shook Jake's extended hand warmly, surprising him with his strong, purposeful grasp.

"I'm Geoff, Geoff Berkeley. I contacted your agency."

"Pleased to meet you," said Jake.

"Let's get out of the sun and have a drink, shall we, Jake?" he added, motioning toward a nearby caravan which was doubling as a cafe selling teas and burgers to people working on the nearby industrial estate, as well as occasional visitors to *U-534*.

"Why the hell not, I mean, that is to say, why not? After you, Padre,"

Jake responded, slightly embarrassed by the faux pas of using one of his favorite phrases.

The two men strolled over to the makeshift café, passing beneath the stern planes of *U-534*, past the massive twin screws of the Type IX U-boat, and past the split plating on her external hull. Evidence of more adventurous days.

She was sunk in May 1945 by bombs from a Royal Air Force Coastal Command Liberator, for which the RAF pilot had earned a Distinguished Flying Cross. *U-534* had lain undisturbed in the silt of Scandinavian waters for almost fifty years until she was awoken from her slumber, having been raised in 1993. She'd finally found a home in what the Beatles LSD-fuelled songs of the sixties had named the land of submarines, Liverpool. *U-534* was no yellow submarine, however. Instead she wore an autumnal rust color, with plenty of evidence of what saltwater can do to steel.

"You know, you don't have to walk on eggshells around me, Jake. You're right; I was indeed a padre in the dim and distant past. Language like that is no surprise to me; I've heard a lot worse. And if a man says what he means, I can assure you it's no sin." The priest laughed.

Jake was relieved at his easy escape from the clumsy comment, but sought to make further amends by outpacing the priest and beating him to the draw to buy the tea, offering a burger, too. The padre declined, pleading having to watch his weight, although he, like Jake, had the lean look of someone whose old military habits extended into the regime of staying in trim.

"My accountant told me you would fill me in on the details, what kind of operation did you have in mind?" Jake asked, as the two men sat upon some discarded upturned pallets on the waste ground near to where the salvaged U-boat now stood.

"Perhaps it would be easier if I started at the beginning," the priest offered.

"Go on," replied Jake, intrigued, but a little wary.

"Okay," the priest breathed, as if limbering up for a long tale. "*U-534* over there is no ordinary U-boat, in more ways than one, Jake. You see, as well as being unique in that she's one of the only ones that's ever been raised, she's got quite an interesting history, which I'm sure you know something of. But I daresay you don't know the whole story."

The priest took a good pull on his cup of tea and continued. "*U-534* made the last patrol of any Kriegsmarine U-boat in World War II, and like a lot of those final U-Boat missions, she wasn't setting out to sink Allied ships. You probably know that many U-boats set off in the last days of the Third Reich. Some were taking jet parts and atomic weapons research to Japan in the hope that the Axis might somehow fight back. Others were sneaking high-ranking Nazis to Argentina. But some were up to something far more important."

"More important than a possible Nazi atomic bomb?"

The priest paused for a second. Jake figured there was clearly something big coming, and since the history of U-boats was one of his chief interests, he was excited to know what it was.

"Well, perhaps you won't think me too dramatic if I said, the fate of man?" the priest replied earnestly.

The comment stopped Jake's easygoing mood dead. This was not what he'd expected to hear. "I'm sorry, the what?" Jake asked, as he wiped a drip of tea from his chin, not really sure if he'd heard the priest correctly.

"Yes, I really did say that. Of course, you might choose not to believe me, but as you'll see, that hopefully won't affect our arrangement."

"Our arrangement? We have an arrangement? You'll forgive me if I point out that I have no idea what the he... I mean, what on Earth you're talking about, Padre."

"Well, oddly enough, Jake, 'hell' might actually come into it. But enough of the mystery, let me carry on with the tale and it'll all become clear."

"Curiouser and curiouser, as Alice said. Okay, Padre, like I say, you got my attention, go on."

"As I was saying, Jake, some U-boats had other cargoes. Mostly just Nazi loot from the war, paintings, statues and the like, but among those things was something we would really like to get our hands on. We, meaning the Roman Catholic Church."

"And that would be?"

"That would be the Lancea Longini," the priest answered.

"A sports car?"

"Yes, I know, it does sound like that, doesn't it? But perhaps you might know it by it's more popular name; the *Spear of Destiny*."

"The Spear of Destiny? Like the Holy Grail--the Spear of Destiny? The Roman spear that killed Jesus Christ at the Crucifixion?" Jake replied, dumfounded.

"Yes, the actual Spear of Destiny," the priest answered with not a hint of doubt.

"Woah, woah, woah, excuse my French, Padre, but you have got to be shitting me. I mean, I thought all those supposedly holy relics were a bunch of mediaeval fakes sold to gullible European rulers by people back from the Crusades? Why would the Church be interested in a fake?" Jake asked.

"Well, yes, you're right, most of them are fakes. But not this one," the priest explained. "Look at it this way, you know that Jesus was a real historical figure, and you also know that he was crucified. And obviously you're well-read in Scripture, you know that he was stabbed with a spear while on the cross. In point of fact by a Roman soldier named Longinus, which is where the fancy Latin name for the Spear comes from. Are you with me so far?"

"Oh, I'm listening. I can't believe what I'm hearing, but I am listening.," replied Jake.

"So, therefore it follows that the spear with which Jesus was stabbed had to exist, yes?"

Jake nodded in agreement.

"So, is it really so hard to believe that it exists still?"

Jake thought for a moment and then spoke firmly: "And I suppose you can prove all that to me, right?"

"Well, not without the Spear itself and an army of archaeologists to hand, not absolutely. But perhaps this will help to persuade you: what do all the battles of World War II in the first three years have in common, Jake?"

"The first three years, let me see, so 1939, 1940 and 1941. That would include, Czechoslovakia, Poland, France, Belgium, Romania, the low countries, France, Norway, the Battle of Britain, North Africa, and Operation Barbarossa of course…the German attack in the East," Jake said slowly as he added them up. "Did I miss any?"

"Well, there's the small matter of Pearl Harbor, several U.S. Pacific bases and all the other Japanese conquests, but if we're talking about purely the Nazis and not including Italy, then yes, basically you've got all the big ones."

"And you want me to tell you what they've all got in common?" Jake asked.

The priest took a sip of his tea, and a deep breath, before he continued. "What they have in common is that without exception the bad guys didn't lose any of them."

Jake thought about that for a second. "Aaah, but that's not true, Padre, the Germans lost the Battle of Britain, and that's well inside your first three years."

"Well, not quite, Jake. You see, the Germans didn't really *lose* the Battle of Britain, they just didn't win it."

Jake let out a quick laugh.

The priest continued, "And that's because they never really wanted to win it, in fact, they never wanted to fight it at all. As you know, Hitler was all for kissing and making up with the British, and lots of the British were keen on that idea, too, having seen how easily the Wehrmacht went through Poland, France etc. If it really hadn't have been for Winston Churchill seeing the fight against Nazism as his crusade against evil, and choosing to go after Hitler, the Battle of Britain probably wouldn't have ever happened. And the world we know today would have been a very different place, with Europe either in Nazi or Communist hands. All Hitler ever wanted to do was go east, but Churchill wouldn't let go of his coat tails.

"And there's something else you might like to know about the Battle of Britain, too, but we'll save that for later." The priest finished his tea and tossed the foam cup a good ten feet into a waste bin, a perfect shot which seemed to punctuate his argument and make it more convincing.

"Okaaay," Jake nodded slowly. "So the Germans never lost a major battle before 1942."

"Yup, and that's a big list of battles, too, isn't it, Jake? I mean, to not lose a single one? Hardly seems like a fluke, does it? But, they did lose something else at that time..."

"And that would be the..." Jake started.

"In 1942 they lost the Spear of Destiny," the priest confirmed in a congratulatory tone, pleased that Jake had guessed that part of the story.

"You see, whether you believe it or not, we in the Church do believe that possession of the Spear granted them some kind of power which made defeat, if not impossible, then much less likely when they possessed it. On the other hand, you might take the view that it's just a lot of superstitious nonsense. But even then you have to admit that looking at some of the Nazi's troop dispositions and equipment levels in many of those early battles, it does seem somewhat miraculous that they managed to pull off the kind of victories they did. Half of them were nothing more than gigantic bluffs. But they still worked."

"Well, good tactics and bluffing are not miracles," Jake countered.

"True, Jake, true. But the fact is, the Nazi's first target was the annexing of Austria, and that was no coincidence, it was so they could get the Spear. The Schatzkammer Imperial Treasury in Austria was where the Spear was kept before the war. Then the Nazis took it back to Berlin, and off they went, winning victory after victory. But in mid-1942 it went missing, we believe someone in the Nazi party took a shine to it, as Nazis were wont to do with archaeological treasures. Nevertheless, they recovered it much later in the war, by which time things were going badly for them. Although they did manage some remarkable comebacks with jet fighters and missiles and the like, even despite round-the-clock bombing of their industry. Just not soon enough to win the immediate fight, so they were intending to take the Spear to South America and regroup for round two, you might say. And they also made something at that time which will be of particular interest to you."

"But, wait a minute," Jake countered, "even I know that the Spear of Destiny was captured by the Allies and returned to Vienna by General Patton. So there goes your nice story, Padre."

"Yes, but you might also know that the Spear returned by Patton was tested by a metallurgist in 2003, and he found that it dated from the seventh century at the earliest."

"Meaning?" Jake asked.

"Meaning it isn't the one the Nazis took from Vienna, but a copy made from another relic," the priest confided in a conspiratorial tone.

"This is all a bit Indiana Jones, isn't it, Padre? It's well known that there are several Spears claiming to be the one around the world, aren't there? I mean even the Vatican claim to have one, don't they?" Jake countered.

"Yes, you mean the Etschmiadzin Lance in the Basilica of Saint Peter, supposedly found by one of the crusading knights in the Holy Land, following a vision he'd apparently had. Trust me, Jake, it's not the real thing. We know that. The real one was put aboard *U-534* over there, with the intention of sailing it all the way to Argentina."

"Yes, but an Allied B-24 Liberator stopped all that. By depth charging her in the Skageraak and sinking her, which hardly fits in with your undefeatable army theory."

"If that's true, then where is it, Jake? It wasn't found on board when they raised her."

"Well, maybe that's because it was never on board her in the first place, Padre, and maybe the one Patton had was the real Vienna Spear, and maybe your story is just a bunch of wish fulfillment from the Pope. Who incidentally, is a German, you might have noticed," Jake answered, slightly more aggressively than he'd intended.

"But you are forgetting one part of the story of *U-534*, Jake. When she sailed, that minutes before she was due to depart, a mysterious passenger was berthed on her. With, I might add, a sealed package that was guarded jealously. And when *U-534* was hit and the crew abandoned her, a boat came along almost immediately, rescued the mystery man from the water and sped him away. Leaving the rest of her crew to await rescue many hours later, by which time three of her crew had died of exposure. Doesn't that seem a little odd to you?"

"A little odd perhaps, yes, and slightly heartless, but it doesn't mean he had the Holy Grail or the Spear of Destiny or any other mythical thing shoved up his jumper, now does it?" Jake concluded.

The two men sat in silent contemplation for a moment, looking at the old U-boat's brown metallic form as it turned golden in the shifting beams of sunlight. Jake wondered briefly if the golden appearance had some sort of significance before he cast aside such a fanciful notion, preferring to keep himself rooted in earthly things that he knew for sure.

"Anyway," he started, "what has all this got to do with me? Sure, I used to be on submarines and I'm interested in this old U-boat, but that doesn't explain why you're talking to me about this. I mean, the Spear, if it ever was on that boat, is long gone. So anything I could tell you about U-boats is no use to you. And besides, there are certainly people who know more about U-boats than me."

"U-boats yes," answered the priest. "But you have other talents which make you a logical choice for what we have in mind."

"I see. So you want me to do something for you?" Jake asked.

"Yes, Jake, we do. We want you to recover the Spear."

"And I would know how to do that because...?"

"Because of your particular skills, Jake," replied the priest.

"Of course, a lot of this hinges on whether I believe your story or not."

"You can choose to either believe it, or choose not to. We don't expect you to convert to Catholicism, Jake. But we do know that you are the man for the job we have in mind, and one thing we also know is that you believe in cash up front. We've spoken to your accountant and we'll sign a deal to retain your rig for eight weeks, plus a $500,000 accomplishment bonus. I take it that is something you can believe in, Jake?"

"Five hundred thousand dollars?" Jake repeated slowly, exhaling. "Padre, I'm a believer. Who do you want me to kill?" Jake laughed. "That was a joke, by the way. What did you have in mind?"

"You know that the Vatican is no pauper, we are certainly good for the cash. So do I take it you are in then? I have to know because what I'm about to reveal cannot be told to anyone who is not with us," continued the priest in a tone that had become deadly serious.

"As long as it doesn't require anything too illegal, or killing anyone who I don't think richly deserves it, then you can consider me in."

"Good, because I think you're going to like this next part, Jake. The submarine connection doesn't end there. That boat which picked up the mystery man from *U-534* took him to rendezvous with another submarine. The Nazis were nothing if not thorough when it came to backup plans for something as important as this was to them. But the submarine they ended up using was largely untested, which is why it was not the first choice with which to transport the Spear. I'm talking about a Type XXI submarine."

"But with the crew of *U-534* possibly about to be captured, the Nazis could not risk taking the Spear to the same destination they had in mind for *U-534*, whose crew might reveal the route, and so they took it to another place, which is where your expertise comes in. The Type XXI went to the South Atlantic, under the ice en route to a facility the Nazis had constructed in the Antarctic. But it never made it there; it was trapped in the ice under the Antarctic. It's still there now, and we believe the Spear is on board," the priest explained.

"Would you mind telling me how the Roman Catholic Church came to know about the movements of a top of the range Nazi submarine when all the Allied intelligence services didn't, Padre?" Jake asked.

"We've amassed a lot of information over the years. You will know it is there when I take you to the location. How I know the location is something you wouldn't believe even if I could tell you, which I can't. This is one thing you're going to have to take on faith, Jake, at least until we get there and you see for yourself."

"So, what you're really interested in is my experience with drilling operations and my knowledge of submarines, huh?" Jake asked. "Okay.

When do we go, and how the hell do we get there?"

"We will need your drilling equipment to be flown to the site, and we would prefer to keep your drilling crew to a minimum, for obvious reasons. Naturally, we assume you can arrange that part of the matter, but we will certainly be able to provide an aircraft suitable to fly your equipment from Alaska, where I think I am right in saying your portable drilling rig is presently located. Having studied your work, we know a C-130 will be adequate for the operation, and capable of transporting your equipment. You'll be pleased to know that I am your liaison for the operation and will be accompanying you on the expedition. You needn't worry, Jake, before I was in my present post with the Vatican, my Army experience included a lot of operations on ice, so I won't be a fifth wheel," advised the priest.

It seemed that the priest knew a lot about him, and his operations. "Did you know I was going to say yes to this caper, Padre?" Jake asked mischievously.

"You know God gave us free will, Jake, so I couldn't possibly know that now, could I?" laughed the priest. "But if I'm honest, as long as you didn't dismiss me as nuts before I'd managed to tell you the whole story, my bet would have been yes."

"That's why I never played poker with the last padre I knew."

⊕   ⊕   ⊕

Four weeks later, the privately registered, ski-equipped C-130 gingerly touched down on the barren Antarctic ice floe. It taxied up to a Spartan camp that would have been difficult to spot from the air, given that the few temporary tent-like buildings were all camouflaged a blinding white color. It was obvious that without a GPS location, no one would ever have been able to find the place, and Jake began to wonder how often the Catholic Church got up to this sort of thing, as it hardly seemed like an amateur operation.

As requested, Jake had kept the crew down to just himself and Steve Palmer, a freelancer he employed from time to time whom he trusted, Palmer having a good deal of experience, and not just in drilling operations.

With just two people to set up the rig and operate it, it would mean a fair bit of hard work, but the pair of them had joked that there would be nothing wrong with working up a sweat in the Antarctic.

Accompanying them on the aircraft had been their contact, Geoff Berkeley, the priest. The only other people on board being the two pilots, who weren't the chatty kind it seemed, but judging by the fact that they were all in one piece and down safe, had certainly done this kind of thing before. Jake had been jokingly referring to them as the 'Salvation Air Force' throughout the flight, but they didn't seem to find it very funny.

Three hours later, the C-130 was unloaded. The padre enthusiastically put his back into helping them remove equipment from the aircraft's cargo hold, along with a man who the priest called Pepe, who had been waiting at the Antarctic site when they arrived. Both the priest and Pepe seemed to be very familiar with the systems of a C-130's cargo loading ramp and its related mechanisms. And on the flight, the priest knew enough about traveling on a C-130 to be aware that sitting over the wings was the best way to avoid airsickness, for which traveling in a C-130 was famous. All of which got Jake wondering what other adventures these two had been on in the past. He doubted this was their first Spear expedition. Pepe looked more Eastern European than his name would have suggested, and had apparently been manning the camp for a few weeks, judging by his beard growth. But Jake judged it best not to ask questions, and doubted he would have gotten truthful answers in any case.

With the aircraft unloaded, the crew wasted no time in firing up the engines and setting off back to Chile. Everyone watched it depart, knowing that their link with civilization and comfort was now gone.

The priest informed Jake that they would radio for a pickup when they had the item—which he did not refer to by name in front of anyone—and they could be off the ice within eight hours of the call, weather permitting. The priest had been as good as his word where the money was concerned; Jake was now considerably richer, minus his expenses and paying Steve. That being the case, the priest would presumably be as good as his word on the pickup too. Which was a good thing, as Jake had never liked the cold, although it seemed they were lucky with the weather, with hardly a breath of wind in the air.

With only two hours of decent light left, Jake and Steve set about rigging the drilling gear up at a location based on the seismic data, radar and magnetometer readings which the padre had provided, apparently another string to his considerably capable bow. Doubting they would have it ready to go before it would be too dark to risk continuing to erect it under floodlights, they actually made good time and managed to get the equipment ready to go quicker than expected. With no wish to spend any more time on the ice than they had to, Jake and Steve arranged to operate the drilling rig in shifts so they could drill down continuously to the entombed submarine. They would be sinking a shaft just wide enough for a man to descend to the wreck, which was apparently in one piece, sitting upright around seventy feet under the hard-packed ice.

With Steve volunteering to take the first shift, Jake joined the priest and Pepe in one of the tents. They had a quick meal, courtesy of Pepe's questionable cookery skills, followed by some much-needed sleep.

What seemed like just minutes later, but was in fact six hours. Steve woke Jake for his turn at minding the rig, informing him that the ice was very tough with some rock in amongst it, and that they had made just twenty-two feet in six hours.

Drilling continued throughout the following day until after twenty-seven hours, they had hit the metal of the Type XXI's hull. Trying to aim the shaft for the hatch on the foredeck of the German submarine, based on blueprints, seismic and radar images, it seemed that they were only about two feet off target. Given the poor resolution of the imaging, it was better than they could have hoped for.

This was where things got tricky, and it was Jake who volunteered to go down the shaft with the air chisel to clear the ice away from the hatch. It was a claustrophobic and dangerous operation, but Jake made good progress and within four hours he had the hatch exposed and free of the glistening bluish-white ice of the chasm.

Time, ice, and the pressure of the shifting floes had certainly not done the U-boat any favors. In addition to being rusty, it was clearly somewhat bent and crushed in places, as evidenced by the many ripples in the steel. This area of tundra wasn't very stable. Nevertheless, this had worked to their advantage, the hatch dogs were apparently almost all sheared off, and the air chisel had it open in less than five minutes.

Before attempting to enter the submarine, it was necessary to retrieve the airline and the chisel back up the shaft. Jake was able to determine over the radio that although curious, nobody else shared his passion for rusty old German submarines enough to want to risk coming down for a look. When all the ancillary equipment was clear, he descended into the frozen sub.

Jake proceeded carefully through the sub's stale air with a powerful flashlight and the charges necessary to blow the safe in the captain's cabin, where they had determined the Spear was most likely to be located.

It was not a pleasant experience.

The sub had remained watertight it seemed, but the air was stale even with the hatch open, and occasionally Jake found himself bumping into one of the submarine's former crew, who were all grotesquely mummified. It appeared some had chosen to kill themselves too rather than suffocate. One or two bodies showed evidence of a self-inflicted shot to the head.

Having been a U.S. Navy SEAL prior to setting up his dangerous drilling operations company, Jake had found himself on military missions to various dangerous places throughout the world, and he had certainly seen his fair share of dead bodies. But like most former soldiers, the experience had left him with a greater appreciation of how precious life was than most people had, and to see these men now like this was a grisly and sobering thought.

As he picked his way through the mess of clutter on the submarine's deck grating, a deep rumble permeated the sub's hull and he felt the vessel shift perceptibly underfoot.

Disturbed by the drilling, the ice floe was shifting.

On the surface all hell was breaking loose. The shaft was collapsing under the pressure of the grinding ice flow and it had filled in to a depth of about halfway up the shaft with lightly packed ice. Steve was frantically calling on the radio to try and raise Jake, who obviously did not know that his sole exit was now well and truly blocked.

Blissfully unaware of the problem, yet aware that time wasn't on his side, Jake hurriedly found the submarine captain's safe and placed the charges in the location he had been advised would yield the best results without damaging the contents of the safe. A piece of advice he was forced to trust, as safe cracking was *not* among the many skills he could count as being experienced in.

Stepping back much further than he had estimated was the safe distance for the shaped charge, Jake thumbed the detonation trigger, remembering to open his jaw to prevent the noise of the blast in such a confined space from permanently redesigning his eardrums.

The bang, although quite small, was nevertheless deafeningly loud in the confines of a sub, and it was with ears ringing that Jake stepped back into the cabin to examine the results. The smell of acrid smoke from the explosion mixed with the stale air was nauseating, and made breathing difficult. Jake's torch beam arced through the smoke from the explosion and found the safe, with the door an inch ajar.

Even though Jake was convinced that the Spear would prove to be nothing more than a rusty old fake—assuming it was even there—it was with some trepidation that he opened the safe. He'd get to keep the money regardless, but he found himself wanting it to be there. As the door swung open with a shower of powdery rust filling the torch beam, inside Jake saw amongst a collection of documents, something wrapped in what looked like muslin cloth. Gingerly he retrieved it, and also stuffed the paperwork he found into his jacket pocket, before opening the cloth binding around the heavy object. And suddenly, there it was in his hands. A spear with years of pitting and wear, it was unmistakably *old*. Of course, that didn't mean it was the genuine article, but there was something about it that Jake couldn't put his finger on, even in the dim, torch-lit surroundings of the old submarine. It seemed to have an aura of power that was, if anything, slightly disturbing. But now was not the time to speculate on such matters.

Suddenly, the boat shifted again and there was an almighty crash. Whatever it was, Jake knew it wasn't good, sliding the Spear into his slim backpack, Jake hurried back to the hatch as fast as he could manage through the smoky torchlight.

When he reached the hatch, his heart sank. It was obvious that the shaft had collapsed, ice was solidly blocking the hatchway, and it was equally apparent that there was a lot of it. He couldn't move it even a millimeter. In vain he tried the radio, but he knew deep down that the transmitter wouldn't get a signal out, surrounded as it was by the ice and steel of the sub's hull.

Up on the surface, things had gotten worse. The edge of the shaft had also collapsed in, canting the drilling rig over and buckling the support stanchions; it was now useless. Even if there had been enough ice surrounding the edges of the shaft to support the rig, it was completely beyond repair with the tools they had to hand. And it was the only way they could drill down to where Jake was trapped. Given a few weeks, they could have used the air chisel to cut their way through if the ice would stop shifting, but by then Jake would obviously be out of air. Forced to retreat back from the collapsing edge of the shaft, the three men looked around in desperation, trying to think of something they could do, and knowing that there was nothing that could be done in time.

It occurred to the priest that maybe nobody was meant to have the Spear, and perhaps that was why this had happened. Regret began to snap at his conscience.

Below, Jake was having trouble staying on his feet as the ice violently shifted the submarine around. He felt the deck tilt over at what seemed an impossible angle, forcing him to grab a wheel on the overhead piping and hang on to prevent himself tumbling back down through the submarine's interior.

The inner hatchways seemed to be below him now, although he knew that this surely could not be so, unless the submarine was sinking into the depths. The idea made him panic for an instant, and then the crazy thought occurred to him that he would now become part of this relic, too. Perhaps even worth something himself someday. A smile creased his lips. There was little he could do but wait for the pressure to crush the hull, and there was no point in crying about it.

Above, the men looked on in amazement as one hundred yards away to their left, the sixty-year-old wreck of the Type XXI broke through the surface of the ice, bow first, looking for all the world as though she was surfacing out of a frozen sea. They had to step back to avoid being showered by huge slabs of ice which crumbled off the submarine's deck and careened along the surface of the ice floe in all directions.

Steve let out a shout as he saw Jake emerge from the hatch of the submarine amid a shower of powdery ice, the sub now poised with her prow pointing thirty degrees into the sky, the bow almost fifty feet above the surface of the ice floe. As the deafening roar of the cracking ice subsided, all three men watched in jaw-dropping silence as Jake cleared the hatch, then slid head-over-heels down the submarine's deck onto the surface of the ice floe. Jake disappeared behind one of the huge slabs of ice, but appeared seconds later, running for all he was worth toward the safety of the solid ice. As he did so, an ear-splitting screech of tortured metal seemed to surround the men, and the submarine ground its way back under the ice to sink out of view.

Breathless, Jake collapsed on the ground in front of the priest. The three men looked on in stunned silence.

Presently, Jake reached into his backpack, fishing out the Spear.

"Is this what you were looking for?" Jake quipped in between fighting for breath.

The priest took the Spear from Jake's outstretched hand and eyed it for a second.

"No, we were after a different Spear. Did you not grab that one too?"

"Very funny. Shouldn't you be blessing that or something, instead of making jokes, Padre?"

Subdued cracking sounds boomed from under the surface of the ice floe, but it seemed the shift was subsiding, and the men noted with relief that the radio shack appeared to have escaped the icy onslaught. Jake thumbed in the direction of the booming noises.

"You see, Padre, I don't think He appreciates your sense of humor."

'Damn it, Jake, I don't know how the hell you made it out of there in one piece! I wouldn't have believed it if I hadn't seen it with my own eyes," said Steve with obvious relief, joining in the laughter whilst shaking his head incredulously.

The priest slipped the Spear into a small bag and secreted it in his own rucksack. He and Pepe walked over toward the radio shack to make the call for the aircraft to return and pick them up.

The priest opened the door to the shack, paused, then turned and shouted over to the two drilling experts, who were now both considerably richer, both in money and unusual experiences.

"So, now do you believe in miracles, Jake?"

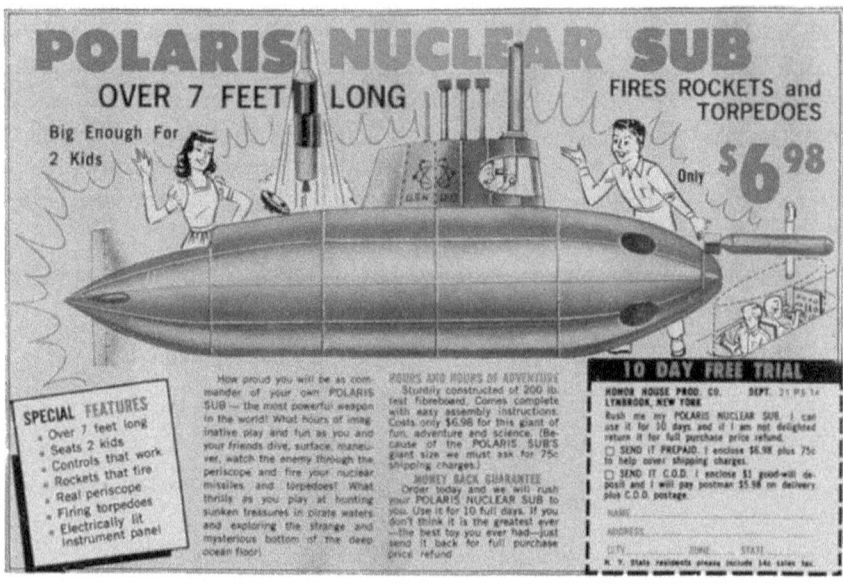

Growl, Tiger

Ron Gorence

The ship passed between Najimo Saki and O Shima lighthouses while it was still dark. When the short-long flashing of Tsurugi Saki's white light was sighted dead ahead, she came around to a northerly course toward Tokyo Bay. Normally, the ship made fifteen knots at standard speed, but the pit-log showed eighteen; the men on the sticks in the Maneuvering Room had been told repeatedly to maintain standard turns, but the screw-count had gradually increased each time. Channel fever was rampant below decks. The showers had run all night, and the diesel-fuel smells in the air mingled with Vaseline Hair Tonic, Mennin Aftershave, and Aqua Velva. Two rocks sticking out of the water at the southern end of Tokyo Wan, called 'The Brothers', became visible at sunrise and the maneuvering watch was stationed. The officer of the deck regained precise control of the ship's speed, and the engine-room snipes, who hadn't seen sunshine for a month, rushed topside in preparation for handling our mooring lines.

The *Razorback* (SS-394) had been on Northern Patrol for thirty-six days, not a record run, but long enough for Baby Huey to develop a hearty thirst to match his size. Upon mooring, the ship's log read: '0830 - Moored starboard side to Berth 1, Yokosuka Naval Facility, Yokosuka, Japan. Present are various units of U.S. Navy and Japanese Maritime Self-Defense Force.'

Huey sprinted across the brow and jumped into the first taxi in the long line of base cabs lined up awaiting our arrival. He was shouting *'Hiako, hiako'* to the driver, who had wanted to wait for a full load. Some crewmembers pretended they were too cool to have channel fever while others attacked the second taxi, but all hands understood the fever well enough to overlook his almost-unforgivable breach of pigboat etiquette, and so Baby Huey was in the Starlight Club by 08:45. It was almost noon before enough crewmembers had assembled to completely obliterate Starlight's early-morning tranquility.

When I finally drifted into the Starlight, everyone but the serious drinkers had a gal on his lap or alongside. These were beautiful women, in western dress like bobby-soxers back home, or in full-dress kimonos. They would have been breathtaking—all of them—even if we hadn't been smelling diesel fumes and each others' armpits for two months. Baby Huey's regular, Mioko, sat across from the bar on a bench and pouted. I guess she realized that tonight her 'big teddy bear' would not be promising to marry her and take her back to Ohio. He was dead-serious drinking. She glared at the bottle that he was stroking with uncharacteristic tenderness.

Huey was about six-four or five, and almost as tall. He weighed 250 or so, with no beer-gut. We drank our suds from Asahi bottles in the Starlight, unlike at picnics, in the days before aluminum cans, where Baby Huey had always crushed his beer cans in one giant hand to improve their throwing ballistics. Now, he brushed imaginary dust from the lip of his Seagram's bottle.

We ate squid on a stick, drank Saki, Jack Daniel's, San Miguel, and anything available while we smooched, smoked cigars and discussed world affairs. Then just about sunset every face, the soberer ones at least, turned toward a commotion at the bar. Huey was half-standing, with the left cheek of his butt on the bar stool, leaning on his elbows toward Papa San across the bar; it was starting.

"Yeashhh, I'll flip you fer the jukebox..."

He slammed a hundred-yen piece on the bar, and with the back of his other hand cleared the stool next to him, where a shipmate, now on the floor, had muttered something like, "No, Babe. Please don't do it."

Everyone calmed down though because Huey lost the toss. He put a hundred yen in the machine, played *She Ain't Got No Yo-Yo*, and a couple of other tunes.

"All yours," he said to Papa San.

At the Starlight, we brought in our own bottles of booze, and Papa San labeled them with our names and stowed them while we were out on patrol or weekly ops playing hide and seek with the surface skimmers and airdales. He charged us a hundred yen (about thirty cents at the time) for a glass of mix and ice and poured a very generous shot. Most of our drinking was hard stuff because Ten High was about a buck, and JW Red was under two dollars at the tax-free package store. The real expense was the cherry drinks for the girls, usually three hundred yen a pop, but if a guy wanted to compete for one of the girls that everyone else liked too, it could go to five or six hundred. These nymphs could drink a torpedo tube dry if anyone had the cash.

Anyway, about the time six hundred yen was starting to sound cheap, and the noise level had gone beyond uproar, Baby Huey was back at it again. He'd won the toss!

He slipped once getting off the barstool, staggered over to the wall behind the jukebox and yanked the cord out of the wall. Dean Martin's *at's Amoré* groaned down to silence like he'd been dropped down #2 periscope well, and Huey put his shoulder behind the Wurlitzer and began pushing it toward the door. Papa San vaulted the bar in one jump, and yelling in Japanese tried to curtail Baby Huey's progress. Papa San was about five feet tall, so I couldn't see him on the other side of the music box; Huey towered over them both.

Baby Huey was bellowing, "I flipped him for the somobishhh, 'n I won," while he brushed off shipmates like flies.

Papa San was shouting Japanese profanities and leaving skid marks on the cement threshold. Huey's eyes were glazed over so bad that he was navigating in darkness like a sub at three hundred feet, but he had little trouble getting the jukebox out the Starlight's door and into the alley which was lined with drinking establishments. He headed it in *Razorback's* direction with vividly colored neon reflections shimmering on its curved glass face, and its little wheels clacking on the cobblestone. Papa San disappeared down Submarine Alley, screaming bloody murder and apparently looking for help.

Knowing that the police or the Shore Patrol would soon be on the way, several of us, like monkeys hopping around a giant organ grinder, tried to talk Baby Huey into putting the machine back, but he was concentrating too hard on overcoming the added weight of a couple of guys on top of the machine.

Finally, after about half a block, the word 'beer' penetrated his fog and he agreed to stop in an adjacent skive house for a large Ashai. He had to take the machine with him, which provoked many excited and unintelligible Japanese voices, but we got him to sit down and drink a beer. By this time we were hoping, in our snickering desperation, that he would pass out so we could just find a skip-loader or something to get him back home.

In fact, he was starting to look a little drowsy when the Shore Patrol

135

showed up, and we were relieved to see him submit docilely to the authorities. He even helped them as best he could by moving one foot in front of the other now and then as the four military policemen dragged him by the armpits to the Shore Patrol wagon.

"You're in a lot of trouble," one of them said as though Baby Huey could have understood English any better than Japanese in that condition.

"Take me t' my f-ing bunk. I'mn tafter-pedo-room," he slurred softly.

"You're goin' to the brig, fella. We got a Status of Forces Agreement here, and you can't get away with that kinda crap."

Of course we were trying to help: we offered everything from a twenty-five-pound can of coffee to a case of steaks for his freedom. Fortunately, Huey kept them too busy to write us up for attempted bribery. Unfortunately though, Huey liked the idea and started mumbling that they ought to let him go with us. Their nasty mood got worse, and they roughly manhandled Baby Huey into the back of the wagon, slammed the expanded metal gate shut with a loud clang, and snapped the padlock shut. Huey sat down, slumped his shoulders and was a teddy bear again. We were just turning away from our hopeless task when we heard the biggest of the SP's bang the side of the cage with his nightstick and bellow, "Now, let me hear you growl, tiger."

Well, that turned out to be a bad mistake. Baby Huey woke up and tried to find the source of the voice through his glazed eyes. He shook his head once and backed up until his butt was against the pick-up's cab. He bent down like a spring compressing and roared as he slammed forward into the locked door.

The door held, but the entire rear wall of the cage, onto which the door was hinged, flew fifteen feet into the street. Huey landed on his knees between the twisted metal and the pick-up. He got up very slowly, with two SP's trying to tackle him, one swinging his baton and one sprinting for the radio in the truck. Baby Huey walked over to the pick-up, grabbed it under the driver's door and tipped it onto its side. A rear-view mirror went flying; sparkling little gems of broken window glass spread out in a fan across the asphalt; it sounded like a dumpster had fallen out of a third story window.

We were screaming, the Japanese audience was screaming, and the SP's were screaming. Only Huey was quiet. Three of the SP's were hauling the fourth out of the pick-up cab with the radio mike dangling from a loose wire in his hand. The Japanese were shouting louder now, yanking on the SP's sleeves and pointing to the puddle of liquid which was growing under the truck. We ran around a corner-bar into another alley, with Baby Huey in tow, and immediately nearly got run over by a cab. He who looks after stray dogs and drunken sailors had provided us with a base taxi.

We got Huey into his bunk in the After Torpedo Room with a minimum of trouble because he was getting kind of tired by that time. I think we were the last ones to get on base before the gate guards started checking all incoming cabs. Some guys complained that they were held up for over an hour. *Razorback* sailors chuckled, but the rest of the Navy apparently was not amused.

Of course, the swift minds in naval law enforcement eventually homed in on *SS-394*, and the skipper had to assure the base commander that the villain would be severely punished at Captain's Mast. It was held two days out at sea after we departed Yokosuka, and Baby Huey was reduced to engineman 3rd class -- suspended for one month. Huey was suspended for most of the three years I spent on *Razorback*.

There was never an engine out of commission when Baby Huey was aboard, not on *Razorback* or any other boat he sailed on. Some said that those diesels were afraid to make Huey mad, but he is the only one I ever saw actually cuddling those 1600-horsepower monsters, and wiping oil from every surface like a mother tending a newborn kid's butt. At sea, he hovered over them, perpetually checking for whooping cough or something. In port, before heading ashore, he always patted them gently on his way to the After Torpedo Room escape hatch. No one had ever seen him actually kiss an engine, but I'd caught him smooching Mioko once, and his scowl convinced me that I should keep my mouth shut for the rest of the century.

I did.

It's hard to pick my favorite subsimming moment, as I have played nearly every subsim since GATO. I always loved the adrenaline rush of coming across a convoy in Silent Service or in the middle of a convoy engagement in Aces of the Deep. My favorite submarine moment was seeing a Kilo class submarine in the Neva River.

Jason "NikiMcBee" Lobo

The Center of the World

Mike Hemming

Today, the center of my world is too large. At one time it was a mere 15x20 feet, and in it we would pack almost thirty men for a movie. Or twenty-four men sitting in contact with each other for a meal, six each at four tables. From the coffeepot to the airtight door leading to the AB berthing area, the mess hall was the center of our world.

It was where an enlisted man went for everything; recreation, knowledge, food, entertainment and believe it not, solitude. Solitude in an area often filled with other men, but when a shipmate got a certain faraway look, he was usually left alone. Unless someone thought he needed some cheering up, and then his 'cheering up' was liable to be at his expense. Abuse about almost anything would be heaped upon the victim until he had to laugh along with the crowd, knowing it was the only way out. A little cruel maybe, but nobody ever cracked up in my times aboard, so I guess our psychiatrist's couch was there too. Right along with eighty other crazed, self-appointed psychiatrists ready to harpoon your self-esteem in the butt until you were as happily loony as they were.

Long discussions on all subjects would evolve through many men coming and going. The length was limited only by the need to set up for the next meal. Day and night, which is a useless measure of time aboard a submarine anyway, men cycle through to go about their lives. On watch, off watch, eat, hit the rack and study quals, a constant churning of men.

It was our doctor's office too. Doc Rohre would set up to do paperwork in one corner with his pipe lit and you knew he had his shingle out. Not that you couldn't stop him or wake him anytime, if you had a serious concern.

Once I was studying quals in the mess hall late one night at sea. Doc was in his corner when a sonarman we called 'Lover' came in and sat down. Lover waited a bit, I think now hoping I would leave, and then started talking to Doc in a low voice. I paid no attention until I heard Doc say in his gravel-voiced not-so-low tones, "Well, when were you exposed?" It was then I realized that you never had any privacy on board. Lover mumbled an answer that I interpreted as five days ago, which sounded about right to me. After all, I had seen all those flicks warning us about trying to spread our pollen among the 'bad' ladies. After three to five days your willy goes AWOL and you die, or something horrid like that. Well, after some more consultation, off they go to give 'Lover' his cure.

His quiet questions became useless when an after battery rat awoke to see Lover's hairy butt being injected with the requisite amount of medication next to his head. That must have been a real eye-popper.

At the next meal the guy comes in and announces for all to hear, "Hey, Lover's got the clap!" Less than three minutes later, the whole boat knew. For the rest of that cruise Lover's arrival in the mess hall would be announced by the sound of applause.

Movies are best shown in the mess hall. I never liked watching a movie in the forward room. Too much room, I guess. Plus, the fact is at sea, the up and down motion on the surface added an extra little something for those disposed to looking for *Ralph*. The junior men got to hold down the projector and the screen. Sitting close enough to a movie screen to hold it down makes for crossed eyeballs trying to focus on the picture.

Mess hall movies had their own extra added attractions and additions, like the hairy-backed EM1 that always got in the picture at the wrong time, looking for a clean coffee cup. Even Ursula Andress' lovely breasts lose a lot when projected on a hairy, sweaty back.

Then there was the duty foul butt that would ease out a silent but deadly stinker in the middle of a reel. The *Carp* had a guy that had apparently invented the stealth fart; it oozed to the far side of the compartment before it started exploding in everyone's nasal passages. Causing them to place the blame for it on everyone else in the mess hall. Added to the body odors of twenty other men far too long away from any sorts of hygiene other that brushing their teeth, it was amazing any of us survived.

Once after a reel was over and the lights turned on, it was discovered that there was no film on the take up reel. Twelve hundred feet of 16mm film in the waterways in a tangled soggy, oily mess. It took some time to straighten that one out.

Between reels, the Coke machine was hit. Actually hit, for if it were done at the right time it would stick and dispense Coke until hit again. The ship's party fund never made much from the Coke machine. Sometimes popcorn was made and ice cream dished out. If you wanted ice cream you always tried to not be the guy to drop down into the freeze box to get it. If you did, you missed half the next reel dishing out ice cream for the rest of the audience.

The mess hall was our game room for acey duecy, bridge, canasta, cribbage and poker. *Carp* was a poker boat. At sea, the table behind the 'pass-through' was the poker table. The metal rim pulled up, a blanket put over the top and rim pushed back down and *voila* -- poker table for six. Seven if the junior man wanted to stand at the end.

Games ran from clean-up from chow to set-up for the next meal. Rules were simple: Bring cash, no borrowing, fifty-cent limit, no kiddy games that took any explanation, cards spoke for themselves. The COB who often played, or just sat and watched, enforced the rules. I never saw anyone that went away from the table busted too bad, just a mite broke. But our main form of recreation was verbal; talking, joking, teasing, lying to or about someone was easy. It took no equipment except a brain and a mouth. Some, however, seemed to leave their brains in another compartment, even then. These verbal jousting tournaments were not for the fainthearted or thin-skinned. If you couldn't take it, stay in your bunk and out of the 'arena of death'.

You could always take heart in the fact that nobody won these verbal 'to the death' battles all the time. If it was decided it was your turn in the barrel, it was best to grin and bear it, and then leave for your next watch, bloody but unbowed.

Every once in a while, I miss that tiny center of what was once my world. A 15x20-foot world almost always filled with men. Men that tried to hide the fact they cared with rough, crude language. A world now only remembered by some crusty old men, tied together by a silver emblem of their belonging, fading slowly away. One by one.

The Diving Alarm Ballet

As I pass between the controllermen, the *oogah, oogah,* "Dive! Dive!" comes over the speakers and they leap to their sticks and rheostats. The engine shut down air lever is hit, rheostats spun down, sticks are thrown, the ballet begins. Generator electricity wanes as the huge storage batteries are called on for power. Sticks pulled to new positions and rheostats spun back up to keep the motors turning. The flurry of intense activity over, minor adjustments made and times logged while listening, always for the sound of water doing something it shouldn't.

As I walk forward at the same time into the engine room, the two men in each one do the shutdown dance. Throttles are slapped down, hydraulic levers pulled to the closed position to shut exhaust valves and drains opened by the throttle man. As his oiler spins the inboard exhaust valves the thirty-two turns to shut it, either the oiler or the throttle man (depending on who is closer) will have yanked the pin holding the great intake air valve open so it falls shut with a loud clang. His inboard exhaust valves shut, the oiler drops below to secure the sea valves that allow the seawater to cool the engines. Then the throttle man checks that everything is secure one more time.

In the control room, the other area of great activity on a dive, lookouts almost free-fall to their diving stations on the bow and stern planes. Quickly the bow planesman rigs out his planes and both he and the stern planesman set their charges to the prescribed angles for the dive. Arriving soon after the planesmen, the OOD, now the diving officer, gives the ordered depth to reach and the angle to do it. Then he checks that all is well and will watch the planesmen to learn if the trim needs changing.

The chief of the watch, having closed the huge main air induction valve, will watch the Christmas tree to see that all hull openings are closed. Then he pulls the vents to flood the main ballast tanks and watches the depth to signal the auxillaryman on the air manifold when to blow negative tank to the mark to arrest our descent into the depths. The manifold operator will hammer open the valve and then close off the roaring rush of compressed air, as needed.

By this time, the trim manifold operator will have arrived from the engine room. After climbing over the stern planesman he will be ready to pump and flood seawater to the tanks. This will trim up the boat to neutral buoyancy.

In the conn, the helmsman will have rung up standard speed so the boat will be driven under by the screws. The QM of the watch will dog the conning tower hatch when the OOD, the last man down from the bridge, pulls the lanyard to close it.

There is no music to guide this dance except calm orders given and acknowledged. Started in a flurry of activity, it will end by winding down quietly to a state of relaxed vigilance by men practiced and confident of themselves and each other. They have done this many times, this graceful and awkward descent into the depths. They do it as fast as is safely possible. This is where they belong, with many feet of sea hiding the strong steel of the hull. Men asleep in bunks half-awakened by the raucous alarm and noisy ballet, drift back to deep sleep, confident they are at home, where they should be.

Fueling and Lubing Around

Smokeboat engines suck fuel like thirsty sailors suck booze after three months at sea. And if you put going home turns on, look out for serious consumption of petroleum products.

Our boats were good for about 11,000 miles on about 110,000 gallons of fuel. Definitely not EPA standards for MPG in our new touchy-feely world. But hell, ours was not a touchy-feely job. We were tasked with being ready to go out and break ships and kill people. So getting about a tenth of a mile per gallon to get there didn't matter to us. That mileage was at a standard bell on two engines. Crank up those big GM's or FM's to full on four and you start using some serious juice. Seems to me about 800 gallons per hour to hit twenty miles per hour. The downside of all this consumption was that it had to be put back aboard. Enter that magical snipe called 'The Fuel King'. It was the only crown I ever wore, except for 'That skinny MM in the AER'.

Fuel king is a rite of passage for any snipe that moves up the ladder in the engine rooms. Kind of like being Spare Parts P.O. except that it is thankless, dirty, cold and you can't steal anything. Going home with buckets of dirty diesel oil will not win you any 'sugar' on the home front. Anyway, the fuel king was any 3rd class snipe that could count to 110,000 without using every toe in the Navy.

Fuel consumption figures had to be kept, and also kept correct so the officers knew how much the boat weighed. Once a trim officer ignored or 'lost' the fact that #1NFO tank was 'empty', i.e. now filled with seawater and that's heavier than fuel. The boat dove about 30,000 pounds heavy.... Talk about a down express. We were in Long Island Sound on school boat ops and stuck our nose in the mud at 125 feet. From the bugged-out eyes of those students in the engine room, I wonder how many went ashore and off to skimmers. After a short discussion with the skipper and the aforementioned officer, showing my numbers were there and correct, I headed aft to the sound of a serious chewing out.

Our fuel, before it could be used, had to be run through a centrifugal purifier to spin out all the dirt and water in it. The purifier had a cylinder that rotated at 25,000 rpm to do this. The water flowed out the end and to the bilge, but the dirt stuck to the sides and had to be cleaned out. With relatively clean fuel at lower speeds the oiler might not have to do it all on a particular watch. But at high speeds with dirty fuel it might be done twice a watch. It was a knuckle-busting and dirty job, even worse in any kind of sea. If you put this monster back together unbalanced or with a bad bearing, it would do a shuddering fandango about the time it hit 25,000 rpm, scaring a newbie oiler's socks off. Me, after hitting the stop button, I waited in maneuvering for things to quiet down.

Fuel from Navy sources was usually pretty good, but if gotten from locals, look out. *Carp* once fueled in the Azores. Any hoses in that place should be inserted in a different orifice and carry a different liquid. Anyway, I don't know what was in that diesel, but it was the hottest burning stuff I ever saw. Percy Turner said it was jet fuel; it burned exhaust valves out at a fast rate. We were replacing a set on a cylinder at least once a day. I couldn't get the engineer to let us shift tanks, the lazy jerk wanted to use fuel from amidships to make his compensation easier to figure. Finally, the old man said shift to another tank because we were running out of spare valves. That damn fuel sat in that tank until we off-loaded it in the yards.

Fueling was a simple process. Snipes hauled aboard and hooked up black greasy hoses and the station pumped it aboard. All the oil king had to do was line up the right tanks and make sure they weren't over filled. If they were, oil went into the water, making them troubled waters. Stories abound of oil kings who learn that sleeping while fueling can cause you to spend days breaking rocks and nights fending off a big bosun's mate named Bubba in Leavenworth. So this one never slept or read skin books while fueling.

Defueling was a little harder and took much longer. The fuel in the tanks was blown ashore with 15-psi air. The problem came after finishing that some tanks were full of seawater. This can't be sent to fuel tanks, so must be blown into a 'doughnut' to separate the oil and water. Trouble was, no doughnut. We waited all night long and it hadn't showed. It's five a.m. and liberty is drawing close. Ah, hell! Let's blow it over the side. At 15-psi, it ain't going fast enough to beat the sun and the oil smell which is getting stronger and stronger. What seemed to be such a good idea to a sleepy mind, now isn't at all. Inching up the pressure to 17-psi and a strong outgoing tide saved the day.

One of the most idiotic things that ever happened with fueling concerned some admiral who was coming to the *Orion*. Somewhere, some moron LCDR decrees whites will be worn on Pier 22 that day. Great if you are some clean finger-nailed QM wandering down the pier to get your frilly undies from the laundry truck. Not so good if you are an E3 wrestling a fuel hose from the pier across another boat. Try as I might, nothing could be done about allowing my men to wear dungarees. Admiral Pantywaist might have the vapors if he saw sailors in—*gasp!*—dungarees. I finally got an extra clothing allowance for my men whose whites were ruined by someone's stupidity. You know, it probably wasn't the admiral's idea but some ass-kissing subordinate. At least, I hope so.

Taking on lube oil was easy here in the States and places like that. But in ports with no naval presence it was a little more difficult. Before *Carp* left for the Med, the COB, an old-time ENCS, calls me up to the messhall, hands me a note and says:

"Take this to shop so-n-so and see Chief Smith, he'll tell you what to do."

"What's this for?"

"An air-powered barrel pump, 'n you'll thank me in the Med."

So off I go, entering the shop, I ask for Chief Smith. I am grilled as to what I want and why I want to see the chief. Sensing the object of my search is within earshot, I say, "Chief Zeigler sent me to see Chief Smith."

Then, from out of a nearby office a voice says, "Come in here, sailor, and tell me why Shorty Zeigler hasn't been shot for impersonating a chief by now."

I go in the office to see the world's oldest ENC. His ribbons look like a billboard on his chest capped off by twin fish and a war patrol pin with a galaxy of stars across the bottom.

"Sit down, boy, how's Shorty?"

Being called 'boy' by this guy was no insult because from the looks of him he had paid his dues when the collection committee was a serious group. Then he proceeds to tell me that Shorty was his hardest case to get squared away back in '40-something, when he came aboard. While talking about what a useless non-qual Shorty was, he reads the note, hands it too a hovering 3rd class, who scoots out like a mouse on a cheese run. Once the chief tells me how hard it was to get the COB qualified, he goes on for some time about the stunts they both pulled on liberty in Pearl and other places. One of which was remarkably similar to one that the COB had recently chastised me for, involving alcohol and returning to the boat not quite on time and in the proper uniform. I am thinking, *Hmm, this is worthwhile info.*

After some time listening to a machinegun delivery of Shorty's numerous crimes and faults, a barrel pump appears beside me like magic. Finishing up, he sends me on my way with, "Say 'Hi' to Shorty for me, and by the way, don't believe a word that sumbitch ever says about me."

At this point it's 1000 hours and I do know enough not to return to *Carp* too soon. So until about a half-hour before liberty goes down, I hole up in Bell's for the day to drink some brews and wonder if Thelma really does belong to the human race. After several, I decide she does, and that story they told about her and Dex in the phone booth is probably not true. Anyway, I can't quite envision the required position as it was described to me.

Returning down the pier, I meet the COB, show him the pump and say, "Chief Smith says 'Hi'."

The COB kind of grins and says, "See you tomorrow."

As I walk away the COB says, "By the way, you can thank me by not believing a word that sumbitch said about me."

"Sure, Chief. Sure."

So it's off to the Med, and along about Naples we need some lube oil. When we pull in, a barge pulls up along side of us and sets ten barrels of lube oil on deck. I got a belowdecks watch to stand, so it's midnight before I can pump the lube aboard. I did have my oiler line everything up, so we are ready to go as soon as I'm off watch. Inserting the barrel pump into a barrel and crack the 225-psi air to it and *kachunk, kachunk,* it's a-pumping. Speeding up the pump to as fast as is reasonable, Bill and I watch it go. As it is midnight and warm in Naples, crewmembers are coming back now.

"How's the beer?" I ask Secor.

"Not bad, I liked the Peroni," he says.

"Well, Bill and I could use a couple. Take this money and go get us a six-pack."

Off he goes, but by the time he reaches the brow he has enough money for a case.

Returning with a case of Peroni's, Bill and I each crack one open. Ahhhh! *Kerchunk, kerchunk* goes the pump, stopping only to move it from one barrel to another. Beers are quaffed by a bunch of snipes, on and off duty. It's taking about a half-hour per barrel, and two barrels for a case of beer. There's always a non-qual to hand money to for another beer run, and life is good. Stories and jokes are told, shipmates are harassed, empty barrels beat on, and oil is pumped.

The *Torsk's* oil king stops by, seeing the barrel pump he wants to borrow it. In unison, six *Carp* snipes say, "Two cases of beer and…um…make it Peroni."

"Or go see Chief Smith and don't believe a word he says about Shorty."

Kachunk, kachunk, kachunk.

The Last One

Out of the morning mist she appears, a still sinister, low black shape. As she does, two men get out of a car and walk to the water's edge. Both are in uniform, the oldest put her in commission and fought a war in her. Then he returned at the end of his career to make her last cruise and decommission her. Retired now, he puts on his uniform once more to say goodbye. All the boats he served on are decommissioned and gone now.

The younger man, his son, qualified on her and sailed on her for four years before returning to civilian life. He too wears his uniform for one last goodbye.

They both watch silently as she slips by, headed downriver to the sea one last time. From this distance, softened by the low fog, she looks well cared for like she was when they sailed her. The man-years they and others spent on her are over now and will only be memories fading away. Father and son salute for the others as she passes abreast of them. The black hull moves on and fades back into the mists of her last journey. Slipping quietly through calm glassy water, she becomes a black ghost of steel.

The father and son hold their salutes as she fades from black to gray to gone in the mist. The tugs mournful foghorn is the last evidence they hear of her. Dropping their arms and turning away, they both catch the sadness in each other's eyes. But as they walk side by side, without words, smiles and twinkles return to their eyes. A memory of shipmates and of times good and bad, but mostly good, return. Memories of her steel, strong and true, that protected her men from the dangers of the deep oceans and her country's enemies. Memories that remind them how much they owe this deep-sea lady.

Because these men remember her, she will not yet die. Because men loved her, she became more than steel, she became alive to them. Because men loved her, will she ever fade completely from men's hearts?

A Cold Wind Blows

There they are, hanging side by side. Warmth and some memories. One is almost new, the other old, stained, with its faded patches and frayed edges. Back then, foul weather jackets came in two styles. The medium green ones with the high, unlined collar and a green lining. It's warm and nice. The other style is the light dull gray green with the shorter collar lined with that brown kinda itchy fake fur stuff. Turned inside out it's great to play Viking in and it's really warm.

The newer of the two, but still aged thirty-some years, was issued to me not long before I got out. When he heard I was getting out the COB said, "Turn that jacket in when you leave."

My other jacket had a busted zipper and was marked with five years of smokeboat snipe living and working. I didn't want to give up the new one. So when the day came I gave a squeeze and a slight twist with two pairs of pliers on the new one's zipper.

"Hey, COB, this jacket's got a screwed-up zipper."

Giving me that look that is issued to all COBs on the boats when confronted with a lying bubblehead, he opens his mouth to say, "Give me the jacket, idiot."

But maybe seeing the look on my face, his voice says, "Okay, keep it... You'll be back."

Sorry, COB, it never happened. It didn't even come close. But I wish you hadn't died not long before our first reunion, so I could have thanked you for the jacket with its quickly repaired zipper and well...a few other things, too.

The other jacket with its truly worn out zipper and patina of grease and stink went with me. The wife never said a word about it. In fact, once before we had it cleaned and re-zippered, I caught her standing at the open closet door with her eyes closed and inhaling.

She is a true smokeboat sailor's wife.

The dry cleaners, however, didn't have the same appreciation for such olfactory stimulation. The lady looked at the jacket lying on the counter like it was made of leper's bandages, and I swear would have picked it up with tongs if they had been handy. But a week later there it was...my jacket faded more now but with just enough black grease stains left. The artwork on the back, my name in red, white and blue inks with the screw drawn under it is still visible. Even the lines indicating motion show. The lines that show a screw backing down, so when asked about it, the answer was,

"Damn straight! I'm backing out of this navy as fast as possible."

The lady said, "That's the best we could do with the stains, and the odor…well."

"That's fine," I said. "It's perfect, just like that."

A raised eyebrow was her only answer. Some will never understand, I guess.

So now when a cold wind blows and I reach for warmth and memories, the old jacket is my first choice. When I walk in it, hands deep in those fuzzy pockets, shoulders hunched to bring the too-short collar up nearer my ears, every once in a while my nose catches a scent of salt tinged air and my feet are back on a slotted walking deck.

I started with simulation games when Sega Genesis produced console games. I had purchased a submarine game for the Sega console (which I still have today) and got hooked on simulation games in general. I did my rounds of flight simulation games like Jane's WWII Fighters, Lock On, B-17 Flying Fortress and IL2. In and amongst these games I purchased SH1/SH2. I did not play SH1 very much as the flightsims kept my attention. I could not get SH2 to run so well on my machine and it always crashed when I would start on my second patrol. I searched around on the web and found Subsim.com had plenty of information of SH2. SHIII came out on the market and although very good out of the box, it did not hold my attention for very long. I dusted off SHIII after it sat for about a year on my desk and went hunting to see how mods progressed and being that I got a clue on how to add mods and play with the files inside the games, back to Subsim.com I went. The people and the site have been so influential I volunteered on the *USS Torsk* restoration efforts taking place.

Chris "AVGWarhawk" Gossweiler

Submarine Dictionary

A

Angles and Dangles - the severe maneuvering a submarine does during training and/or diving/surfacing.

B

BDNW - broke down, no workee workee. An item that is not working.

Blow and Go - to initiate an emergency surface using high-pressure air.

Blowing Sans - or blowing sanitaries. The act of blowing the contents of the sanitary tanks to sea.

Broach - to pop up out of the water while at periscope depth.

Bubblehead - a submarine sailor.

Bug Juice - a generic, U.S. Navy issued, instant drink mix. Origin of name unknown.

Burn a Flick - to watch a movie.

C

Check Valve - a selfish person as in a one-way check valve which only allows fluid to travel one way.

Chop - the Supply Officer, so named because the U.S. Navy Supply Corps Officer symbol looks like a pork chop.

COB - the Chief Of the Boat; the senior enlisted man embarked.

Collision Mat - pancakes.

Crank - to work in the kitchen.

Creamed Foreskins on Toast - creamed chipped beef served on toast.

D

Dancing With The One-Eyed Lady - using any of the two periscopes.

Deck - the floor.

Deep Six - means six fathoms deep (36 feet). Used as a slang expression for throwing something overboard. "We gave it the deep six."

Dink - a sailor who is delinquent in his submarine warfare or watchstation certifications.

Dolphins - the name given to the enlisted or officer Submarine Warfare pin.

Dynamited Chicken - chicken a la king.

E

EB Green - duct tape used by nukes as provided by Electric Boat and reputedly good to 'test depth'.

Emergency Blow - see 'Blow and Go'.

F

Five by Five - loud and clear.

Flapper King - someone who blows sanitaries on himself.

G

Goat - a chief petty officer.

Goat Locker - chief petty officer's berthing.

H

Hacker - an extremely bad movie.

Halfway Night - the only recognized holiday during patrol. This is the day when the patrol is halfway over.

Head - bathroom.

Hollywood - a long shower with plenty of hot water.

Hull Down - description of a ship beyond the horizon with only its masts or superstructure visible.

I

Italian Sound Mounts - veal Parmesan.

Ivan - the Russians.

J

J.O. - a junior officer.

Johnny Cash - dark blue, working uniform so named due to the black pants/shirt. The female variant is known as the 'Eva Braun'.

K

Kettle - the nuclear reactor as in 'teakettle'.

L

Laundry Queen - a junior sailor, usually non-qualified (especially if Dink) assigned to handle his department's laundry chores.

Lounge Lizard - someone who is constantly in the ship's lounge watching movies.

M

Mail Buoy - imaginary place in the middle of the ocean where mail is delivered to all submarines.

Monkey Dicks - Vienna sausages, sweet pickles or any form of sausage. Also known as 'Poodle/Puppy Peckers'.

Mung - any undesirable growth like mold or dirt or any build up.

Mystery Meat - a processed meat that is served consisting of predominantly pork. It is 'advertised' as different things like pork roast, etc, and the only way to tell is by the side dish. So if it's served with cranberry sauce it's one thing, apple sauce, another. Hence the mystery'.

N

No Load - an individual who contributes nothing, as in a circuit that holds no juice.

Nub - non-usable body; someone not qualified in submarines.

O

Overhead - the ceiling.

P

Pegged - reached a maximum as in 'my temper is pegged'.

Pocket Rocket - deterrent patrol pin with a gold star for each patrol made.

Poop-Suit - blue coveralls worn by submariners.

Port - the left side.

Puss Rocket - a hotdog or other sausage.

Q

R

Rack - your bunk or bed.

Rack Back - someone who is constantly sleeping; often combined with 'no load' as in: "he's a no-load rack back."

Rack Burn - the marks on your face left from your blanket or pillow; a dead giveaway that you've been sleeping.

Rain Locker - the shower.

Red-Tagged - not working; so named by the red tags put on broken equipment.

Rim - to rub 'something' along the inside rim of a coffee cup belonging to someone you don't like.

Rollers - hot dogs.

S

Sans - the sanitary tanks. Where the toilet water goes.

SCRAM – 'Super Critical Reactor Axe Man'. This dates back to the early, early days of nuclear reactors when a man (with an axe) would stand by to cut the ropes that keep the rods out of the middle. If an emergency came up requiring the reactor to be shut down, the 'Axe Man' would cut the ropes. This term has survived as a term for shutting down the reactor.

Sheriff - the commanding officer, so named due to the Command At Sea badge, which is a gold star.

Skimmer - a surface sailor.

Slider - a hamburger, so named because they slide in grease.

Split Tail - a female.

Starboard - the right side.

Suck Rubber - to wear an emergency air-breathing device.

T

Target - any surface ship.

TDU - trash disposal unit.

The Man Who Sleeps Alone - either the commanding officer or executive officer due to having separate staterooms.

Three M (3M) Coordinator - someone whose main activities include Movies, Meals and the Mattress. See also: 'No Load'.

Trim Party - a group of several crewmembers who march from the front of the sub to the back of the sub to make the angle severe.

U

Underway - moving out to sea.

Usta-Fish - term for a previous submarine as in "this is how we did it on usta-fish..."

V

Victor - a visual sighting.

W

Water Slug - firing the torpedo tubes with nothing in them.

X

X Division - generic name for the group of non-rated seaman tasked with everything from painting the boat to handling lines.

X-mas Tree - an electronic indicator in the control room. This device shows by means of red and green lights (hence its name) whether hull openings and tank vents are open or closed.

Y

Z

Zero - any officer; as in the 'O' or 'zero' that starts the word Officer.

Zulu Time - just another thing to mess with your brain.

Silent Service: Specialized Submariner Speech from WWII to Present

Tammy L. Goss

Introduction

"It's rack time."

"Honey, I'm pretty sure the washing machine is FUBAR."

"I'm down to the short strokes at work. I should be home soon."

"We don't have any geedunk in the house!"

What? It took a while for me to get used to my husband's special phrases that he uses at home. For the first few years we were married I had to ask him to please speak 'civilian'. Although he has been off of active submarine duty for almost ten years, he still retained some of the special expressions that were an integral part of his job aboard the submarine, USS *Louisville* SSN724. When the time came for me to conduct a linguistic research project, I found a chance to delineate, define, and describe the colorful language of submariners and issue a follow-up of Ervin J. Gaines' *Talking under Water: Speech in Submarine.*[1]

The lexicons of the military are unique sets of jargon and slang that enable the men and women of the military to do their very difficult jobs. Each segment of the military; Army, Navy, Air Force and Marines, has a highly specialized vocabulary that allow for quick and accurate discourse during times of stress and difficulty. While all the major groups have a common core vocabulary, there are subdivisions that allow an even more

---

[1] This article may be found in the scholastic journal *American Speech*, Vol. 23, No. I (Feb., 1948), 36-38.

precise vocabulary for groups such as submariners. This research proposes to highlight some of the specialized lexical items for submariners who served during the period of World War II until present day. Identification will include such information as semantics and semantic shifts, jargon and slang, etymology, usage, and phraseology. Military organizations, like nearly all large, exclusive organizations, develop slang as a means of self-identification.

Brief History

Submarines have been a part of the history of the United States since the days of the American Revolution. On September 7, 1776, the one-man submarine *Turtle* was unsuccessful in an attempt to attach a torpedo to the hull of the HMS *Eagle* anchored off New York Harbor. April 11, 1900 marks the official birth date of the U.S. Navy's Submarine Force, when John P. Holland sold his internal combustion, gasoline-powered submarine, *Holland VI*, to the Navy (Chief of Naval Operations, Submarine Warfare Division). From that moment on, submarines have been an indispensable force in the United States Navy.

The submarine service is an elite force; not everyone is cut out for the rigorous lifestyle of living aboard a submarine, most often for months at a time with over a hundred other men.[2] Submariners are some of the most highly trained people in the Navy and their jobs are extremely technical. Each submariner has his own specialty, but regardless of their job they are required to learn how everything on the ship works in order to become *qualified* submariners. Qualification earns them the right to wear the coveted gold (officers) or silver (enlisted) Dolphin pin.

Because the ship is so compact, with many different personalities on board, life aboard a submarine can be challenging. Many of the crew will build strong friendships that may last a lifetime, and there can be a strong sense of brotherhood, with all the ups and downs that true familial brothers experience. This brotherhood can be seen in the stories that submariners tell, the books that they write, and especially in the language that they use.

---

[2] Although women are allowed in the U.S. Navy, there are no women allowed to serve aboard a submarine. Generally the reason is there are extremely limited living quarters, there is no room aboard a submarine for the separate berthing units and lavatories that would be required for women sailors.

Methodology

This research involved interviewing, via an emailed questionnaire, a number of submariners from a cross-section of time periods. These time periods were roughly segmented into three groups; World War II, the Cold War, and present day. Responses were nicely varied from various rankings of submariners and the respondents consisted of retired submariners and those currently serving aboard a submarine. Response to the questionnaire was immense, with a total of 148 sailors submitting at least one term or definition for the project.

Results and Analysis

Several interesting patterns emerged from this research. Regardless of boat, age, rate, or rank, most terms were consistent in definition. The overwhelming majority of submariners agreed on the terminology, even among those whose service was many years ago or those who were aboard different types of submarines. Additionally, it appears that very few of the words changed from the WWII years until present day, and those that have mainly apply to new technology and not general terms. In fact, some of the terms can be found to date back to the early days of sailing.

General Naval Terms with Historical Maritime Origins

There were several words that are Navy-wide that, appropriately, can have their etymology traced back to several hundreds years in the past. One example of this would be the term 'goat locker'. Goat locker applies to the Chiefs' Quarters and Mess. According to one source: "The term originated during the time of wooden ships when chiefs were in charge of the milk goats on board." Today, the term 'goat' is also used as a term of respect for those aboard who are older, though two former submariners stated that 'goat' referred to a derogatory nickname for a Chief; usually implying a 'yes' man. These definitions were overwhelmingly upheld by several other respondents and were observed being used during the PBS documentary, *Steel Boats, Iron Men*. According to the website *The Goat Locker*, the phrase originated due to a mascot issues between the Army and the Navy.

Entertainment on liberty took many forms, mostly depending on the coast and opportunity. One incident, which became tradition, was at a Navy-Army football game. In early sailing years, livestock would travel on ships, providing the crew the fresh milk, meats, and eggs as well as serving as ships' mascots. One pet, a goat named El Cid (meaning chief) was the mascot aboard the *USS New York*. When its crew attended the fourth Navy-Army football game in 1893, they took El Cid to the game, which resulted in the West Pointers losing. El Cid (The Chief) was offered shore duty at Annapolis and became the Navy's mascot. This is believed to be the source of the old Navy term, 'Goat Locker' (www.goatlocker.org).

Another term that can be traced back to the days of wooden sailing ships is the word 'head', which is used for the latrine aboard ships. According to the Oxford English Dictionary, this word was originally used to denote the fore part of the ship, the bows, and was first seen in use in 1485. Because the latrine was located in the bow of the ship, the use of 'head' for lavatory came into evidence in 1748. This term is still in use Navy-wide, and can be seen in such examples as 'I have to use the head', and 'head break' which is used as a verb, as in 'to take over the watch for someone so they can go to the bathroom'. Submariners are full of interesting and funny stories and the following is a good example of the use of the words 'head' and 'head call' among other terms such as 'rig for red', CO, and the opening line of all submariner stories, 'this is a no shitter…'

> This is a no shitter…. The boat was at PD at night and rigged for black. The FT of the watch requests permission to make a head call. The OOD denies him permission. A few minutes later L____ asks again, and is again denied. The CO is on the conn and orders the OOD to take her deep. Everything needing to be done at periscope depth was done. The OOD orders the boat down and rigs for red. The CO looks over by the console and there in all its glory is L__'s 'torpedo' ummm, shooting off. The CO asks, "L____, what are you doing?" Of course he was relieving himself. He was then given his head call. He said he did it 'cause the OOD wouldn't let him go to the head and it needed to be done!

One term that took root with aviators and submariners, and then spread to the general Navy, is the rather recent linguistic development of the word 'khaki'. Khaki has its etymology in the Persian language where it means dusty or khak (dust). The British Army originally used the material, which has been dyed a dusty/dull brownish yellow color, for field uniforms starting in 1857, during India and Afghanistan campaigns. Khakis were then approved for use by Navy aviators in 1912 and were adopted for submarines in 1931. Ten years later the Navy approved khakis for wear by senior officers in the general Navy, those who are E-7 or above. The word khakis is now often used as KCB (Khaki Clad Bastard) or collectively as

The Khaki Clan by enlisted sailors, showing an adjectival shift to pejoration, in reference to those officers who were unfriendly or abused their powers.

Specific Submariner Terminology

Most terminology that can be categorized as Specific Submariner Terminology is applicable to the technological advances that are required for submarines, such as 'Angles and Dangles', 'baffles', 'boomer', 'Crazy Ivan', and the phrase 'Hot, Straight and Normal'. However, there are several exceptions such as 'Dolphins', 'getting one's Dolphins', 'tacking on the Dolphins', and 'Silent Service', which are more oriented toward the culture of belonging to the Submarine Service than the actual technology.

One of the proudest moments for any submariner is when they receive their Dolphins. Dolphins are the warfare insignia of the submarine fleet and represented as two Pacific dolphins (Dorado or Mahi-Mahi) flanking the prow of a WWII-type submarine cruising on the surface with bow planes rigged out. Dolphins come as gold for officers and silver for enlisted men. Dolphins are earned through a process of qualifying, in which individuals must learn the location of equipment, operation of systems, damage control procedures and have a general knowledge of operational characteristics of their boat. The origin of the U.S. Navy's Submarine Service Insignia dates back to 1923 when Captain Ernest J. King, USN, suggested that a distinguishing device for qualified submariners be adopted. According to Sub Stories: "The Officer's Insignia is a gold-plated metal pin, worn centered above the left breast pocket and above the ribbons or medals. Enlisted men wore the insignia embroidered in silk...this was sewn on the outside of the right sleeve, midway between the wrist and elbow. In mid-1947 the embroidered device shifted from the sleeve of the enlisted men's jumper to above the left breast pocket. Subsequently, silver metal Dolphins were approved for enlisted men."

Getting one's Dolphins means achieving the status of a qualified submariner. This is accompanied by the current, unapproved practice of 'tacking on'. Tacking on was originally used as a rather innocuous term. During early WWII, when a submariner received qualification, they received the Dolphin patch. Tacking on then signified the sewing on of the Dolphin patch, although today the term has shifted to a description of a more ritual group activity rather than a solitary one.

Tacking on as a ritual was officially disapproved of during President Clinton's first term in the early 1990s, though it is still unofficially practiced and a great source of pride. According to one source, the turning point for condemning the traditional rite of passage was the death of a submariner aboard the *USS Los Angeles*. The sailor in question did not die from the ritual tacking on the Dolphins, but allegedly committed suicide after an investigation that was conducted by the chief of the boat who placed pressure on the man to inform on his peers.

There are two types of tacking on; one is tacking on the 'Crows', which refers to the practice of punching the arm of a newly promoted petty officer, also now in disfavor due to past abuses. The other variant is tacking on the Dolphins, which is similar, but instead of the petty officer insignia on the shoulder, it is the punching of the submarine Dolphins pin into one's chest, usually leaving two blood marks on the uniform shirt from the pins on the back of the Dolphins. These bloody marks are a great source of pride, and one submariner states, "…we went out to celebrate. We went to a bar where my Dolphins were taken from me and dropped into a glass containing alcohol. I had to chug it and catch my Dolphins between my teeth before I was allowed to put them back on my uniform. I don't even remember how many glasses I had to drink before I caught them." The ritual hazing of tacking on has changed over the years since submariners during WWII, but by most accounts, there usually is a ritual. When these men received their patch, tacking on meant that they were thrown overboard at the first dock:

> Coming back I qualified on the *Atule*. A big day in my life. I think all submariners are very proud of the day you qualify. The skipper shakes your hand. The first dock you hit, they throw you overboard. It's kind of fun. But I was just tickled because I had worked hard at it and was very proud and pleased to be a member of the crew.

One WWII veteran said in his case there was no ceremony or ritual for the awarding of the Dolphins; he was just handed the patch. He had to sew them on his uniform himself, but it was still a significant moment for him. No matter how much the term and ritual ceremony has changed, tacking on is a great source of pride for submariners for it means they are officially now an elite member of the Silent Service.

According to several sources, the label 'Silent Service' appears to have undergone a semantic shift in connotation, from one of pejoration to amelioration. However, as one captain stated, "*Silent Service* came into use in WWII because any mention of our submarines or their operations was likely to be of use by the Japanese and do harm to us. I never knew it to be a negative term, but always considered it to be a proud description."

Contrasting with this statement are the comments of a WWII veteran.

One of the respondents passed on the information that during WWII it was decided by the military that a newspaper editor should create a sexy name to make submarine service sound more elite in news stories. He then coined the term 'Silent Service' and put it into use in the newspaper stories and newsreels that were shown in the movie theaters. The label was a source of ridicule by all the submariners he knew. In Ervin J. Gaines' article *Talking Under Water: Speech in Submarines*, there is verification of this definition. "A publicity agent's glamorous name for the submarine service. Scorned by submariners, it is not used anywhere in the Navy." Subsequent research has failed to turn up the publicity agent's name or the article in question.

As one veteran put it, "We had been getting the reports the Japanese were making about the submarines of ours that had been sunk by their forces. Tokyo Rose would tell all about submarines they had sunk and we learned very early that the best thing we could do was keep our mouths shut and say nothing about any losses. I think that was one of the reasons we got to be called the 'Silent Service' because we wouldn't hold interviews and tell much of anything about actions of submarines or how they made out."

Although these may be different definitions for the same term, they all have something in common -- Silent Service is now considered an elite section of the U.S. Navy and the men who belong to her are full of pride in the fact that not everyone can be a member of the Silent Service.

Personal and Personnel Terminology

Military slang is used to reinforce the sometimes-friendly inter-service rivalries. Some of these may be considered derogatory and attempts have been made to eliminate them; however, these have failed because it appears that most service members take a certain pleasure in the sense of a shared hardship which the nickname implies. The compact environment of a submarine and the long time the crew spends submerged ensures that a specialized lexicon will develop to help commanding officers and crewmembers maintain control, increase teamwork, and keep undesirable personal and work traits to a minimum. This is often seen in the derisive terms that are applied to those who are seen as not pulling their weight, creating strife, or other objectionable behaviors. These include such names as 'check valve' for a person who is out for himself and doesn't help others, a 'dink' or 'dink bitch' is someone who is delinquent in qualification points, 'KCB' or 'Khaki Clad Bastard', which denotes the higher-up officers. 'Mustang' is used to point out someone who was enlisted but has gone up through the ranks to become an officer and can be used both in a positive and a negative manner, depending on the personal attributes of the officer.

'NUB' is an acronym for 'Non-Useful Body' and is applied to any person who is new or considered inept; literally a person who is wasting precious air and space in a place where personal space is already nonexistent. 'Skimmer' is a term applied to any Navy personnel who is on a surface ship. 'Skimmer' is frequently modified to indicate disgust by the adjective 'f---king', thereby showing the sense of elitism that submariners have in their post. However, this is not limited to submariners only; surface personnel also have a label they apply to submariners — 'Bubblehead'. This is also frequently modified by the same adjective that is used with 'skimmer'. A more interesting phrase is the term 'swinging dick'. This is used to address a group of crewmen and is generally used in an emphatic address such as, "I want every swinging dick working on it," and indicative of the linguistic creativity that accompanies being a submariner.

Consistent Terminology from WWII to Present Day – General Navy and Submariner

'Geedunk' or 'gedunk', however one spells it, is a unique name used by sailors and has consistently been in use since at least WWII. The term 'gedunk' was originally a noun that was used for desserts, junk food, or candy. Later the word generalized and broadened to a more inclusive adjective that also meant any work that was easy, extras, benefits, awards, ribbons, or medals. It appears that almost anything that was 'sweet' or 'easy' could be tagged with the descriptor 'gedunk'.

'Gedunk' is one of those lexical items that have several different historical origins; each one seems plausible, but all are unverifiable. One source states that this came from the sound the mobile candy/food cart had when was rolled, a type of 'geh-DUNK', a rhythmic thumping sound. Another said it was from a cartoon strip that had a candy store named 'The Gedunk' and a third said it was from the German word *getunk*, which loosely means to repeatedly dunk stale bread into coffee to soften it.

Not always is the term 'gedunk' a positive modifier. The usage of 'gedunk' medal is always in a derisive or sarcastic tone, meaning the National Defense Service Medal, which is considered a meaningless medal. Each war has its own national defense service medal and is considered gedunk because you only have to be in the military to get it, even if you haven't directly participated in the war effort. If a sailor has been in the military since Vietnam, he would have three National Defense Service Medals, one for Vietnam, and two for each of the Gulf Wars.

'FUBAR' is an acronym that stands for 'F---ed Up Beyond All

Recognition'. This term was made popular in WWII and has been in consistent use since. Generally it is applied to machinery and situations that seem to have no positive conclusion in sight. Currently, it has crossed over into computer programming language to mean 'failed UniBus address register', though it is unsure whether this is a coincidence or not.

Other terms that have remained consistent from WWII until the present are 'Christmas tree', 'chop' or 'pork chop', 'ladder chancer', 'rig for red', 'bottom', 'crack the hatch', and 'hold me up', among many others. In general, most words that were popular during the early forties are still in use today, and most of the newer versions are related to the new technology that is aboard submarines.

Conclusions

Overall, the lexical vocabulary of submariners has changed little since the 1940s as evidenced by a comparison to Ervin J. Gaines' article *Talking Under Water: Speech in Submarines*. Ervin lists forty-six terms and definitions. All but eight of them are still in use on today's submarines with exactly the same or very similar etymology. More interesting detail is evidenced when a comparison is made between Ervin's categories of official and unofficial Navy terminology. Of the fourteen terms in the official category, all but one is still in use. In the unofficial category, eight of the expressions are now considered to be official Navy terminology. In comparing this research's questionnaire to Gaines' article, the majority of the changes in words, especially in etymology, have to do with the new technologies now aboard submarines as compared to WWII.

Submarine jargon and slang is as rich and varied as the history of the boats themselves. Some lexical items from specific occupations always make their way into everyday language, and this is no less true for the military. A quick glance at the Oxford English Dictionary lists many such generic military terms that have crossed into everyday English. However, according to the article *New Words: Where do They Come From and Where do They Go?*, there is a possibility that the migration of words from subcultures to dominant cultures depends on the interaction between the two cultures. Because the Submarine Service is such a tight, cohesive network of elite brothers, there is little chance that many of the specialized terminology, technical or otherwise, will cross over into everyday speech. It seems for now, the Silent Service will remain silent in sharing its lexicon.

A Sub and Crew Worthy of the Name Texas

Neal Stevens

August 28, 2006 - On her way from Norfolk to Galveston the new submarine *Texas* stopped at Cape Canaveral, Florida to flex her muscles for the media. As she departed the pier for a three-hour transit to deep water, the crew guided reporters through the different sections, displaying a thorough understanding of their ship, her capabilities, and their responsibilities.

Built by Newport News in partnership with Electric Boat Company, *Texas* is the second of the new *Virginia* class of fast attack subs. A host of technical innovations unlike anything on submarines before her make *Texas* unique. The sub is powered by an advanced nuclear reactor that will not need refueling over the ship's thirty-three-year lifecycle. A pilot and co-pilot, who manage the sub's steering and diving control with joysticks, replace the familiar helmsman and planesman stations with their aircraft-like yokes.

The 'sonar gang' has been integrated into a control room festooned with over forty monitors. Listening through the sonar operator's headphones one can hear the clear sound of a diesel engine. So clear, in fact, that it seems to be right outside the hull instead of several miles away. "In many cases our operators can tell what kind of diesel it is, what make, how many cylinders, and if it needs a ring job," says the sonar supervisor with a bit of humor.

And most remarkably—there are no periscopes in the control room. The polished poles that have been a staple in submarine movies have been replaced with sail-mounted photonic masts that do not protrude into the sub's pressure hull. The cameras beam a high-resolution image onto a 30-inch monitor. The age-old command 'Up periscope' may be replaced by

'Raise the photonic mast'.

"After twenty years of looping my arm around the periscope handles, I have to make an adjustment. This control room is more advanced and allows more efficient tactical communication. All that's missing is a captain's chair out of Star Trek," remarked Captain John Litherland.

All this advanced technology requires the brightest and most motivated sailors the Navy can provide. "I have over 300 'clients' to look after," explains Petty Officer Michael Granito, an IT technician who has been aboard *Texas* since November 2002. "All the sub's laptops, desktops, monitors, and around eleven non-tactical servers. Each sonar console has its own server." He deftly touches a screen to illustrate how individual servers are connected within the sub. The level of complexity is staggering yet "we've logged 98% uptime," says Granito.

With a crew of aces and top-shelf equipment, *Texas* stands ready to begin tough assignments. The *Virginia* class has been engineered to match the *Seawolf* class in stealth and open-ocean warfare, but that's yesterday's war. The new mission in the war against terror is to fight and gather intelligence in 'brown water' (shallow coastal waters in hostile regions).

Capt. Litherland says, "The control surfaces and the position of the sail help the hydrodynamics, and the hovering system gives *Texas* superb close-in capability." Texas has a complex ballast control system that literally micro-manages the sub's buoyancy, enabling *Texas* to park in shallow water right off the coast of any hot spot in the world. This also augments another *Texas* strength—deploying Special Forces.

Some aspects of life aboard submarines haven't changed. The living space is confined, forcing sailors to squeeze by each other in passageways. There are four restroom facilities and a single washer and dryer for 140 men.

When the sub reaches deep water, the bridge is secured and the sub is made ready to submerge. The familiar dive alarm sounds throughout the ship. The pilot eases the 7800-ton vessel below the waves. Soon, the surrounding sea pictured on the monitor is obscured by bubbles and foam as the scope dips under the water. The gentle rocking motion disappears, replaced by a silent, steady sensation. Being on a sub underwater is much like sitting in an office building.

That's about to change. After careful scrutiny of the surrounding waters, the crew of the *Texas* performs a series of maneuvers known as 'angles and dangles'. Beginning at 150 feet, the pilot pushes his joystick forward and the sub noses down to 650 feet at a steep twenty-five-degree angle. The undersea warriors in the control room lean away from the slanting deck, still performing their tasks without interruption.

The overhead speaker announces a simulated torpedo launch. In the torpedo room, Petty Officer Monk carefully coordinates the process with the fire control operator in the control room. *Texas* and other modern U.S. subs use wire-guided torpedoes.

"The torpedo will go out there and acquire the target, and when it does, it tells us where the contact is," explains Cdr. Jim Gray. "That's where the wire comes into play." Cdr. Gray is at ease explaining the intricacies of the fire control systems, with good reason. Cdr. Gray is slated to be the next skipper of the *Texas*, taking over command September 20, 2007.

Of the 140-man crew abroad *Texas*, twenty are Texans, including Al Onley, the executive officer. "*Texas* is pretty big, I'm from Greenville so it's a long haul for my family to attend the commissioning, but I don't think anything could keep them away."

Al Onley's family will have a lot of company. The September 9 commissioning ceremony is expected to draw over 10,000 guests and visitors to Galveston, including ship sponsor First Lady Laura Bush, Senators Kay Bailey Hutchison and John Cornyn. Together they will witness a historic event, the commissioning of the most powerful ship to bear the name *Texas*.

Texas Sailors at Liberty to Care

They brought the submarine *Texas* through four years of construction, months of exacting sea trials, and finally, on Sept. 4, 2007, to Galveston for commissioning Saturday, September 9. Almost as soon as the ship touched the pier, the crew of the Navy's latest sub has been engaged in a whirlwind of community activities that would make even the most civic-minded person wilt.

On a warm Tuesday morning fifteen members of the crew put on their work clothes to plant cord grass in the Pierce Marsh Preserve. The Pierce Marsh is a 1,361-acre coastal estuary in Hitchcock. Volunteers use dredged material from developed subdivisions to create marsh terraces at the necessary elevation for inter-tidal plant species to thrive. "We transported them to the area by airboat," said Galveston Bay Foundation President Bob Stokes. "They are a good bunch of guys, I think they had fun."

Evening found *Texas* sailors touring the only other existing ship named after the Lone Star State, the Battleship *Texas* in nearby LaPorte. They concluded their visit with a wreath-laying ceremony.

As dawn broke Wednesday, seventeen *Texas* crewmen held a field day in *Seawolf* Park on the WWII submarine *USS Cavalla*. "We're out here to support the submarine community and help restore this proud warship to its original state," said Senior Chief Richard Lattimer.

The *Texas* sub crossed the bow of the *Cavalla* upon arrival Monday. "When we came in around the last turn we could see *Cavalla* facing out and welcoming us into the port," he said.

Galveston's welcome for the submarine *Texas* has made a favorable impression on the crew. It has been a cheerful homecoming for the Texans on board the sub, such as Senior Chief Marty Ledesma of Lubbock. "We've had a good time. I toured the *Cavalla* about a year ago," he said, while making his way through the confined spaces between the sub's pressure hull and deck to help disassemble old air ducts. "Obviously, if you've been aboard our submarine, the *Texas*, you can tell there's a big difference. But our mission is still the same."

U.S. submarines were originally submersible boats that were most effective against commerce ships. After the war they evolved into stealthy hunter-killers that can stay submerged for several months. "*Cavalla* was one of six WWII subs that the Navy converted to what they called the Guppy class," said John McMichael to the gathered *Texas* crew. McMichael was the chief of the boat on submarines and present curator of the naval display in *Seawolf* Park crew. The bow of *Cavalla* was reconfigured to house hydrophones that listen and detect enemy subs at great range. "*Cavalla* was a predecessor to the modern subs like the *Texas*."

In addition to the many community activities, the officers and crew have to pull their normal duties as well as show guests and VIPs through the sub. "This is the first time I've been in civilian clothes this week," said Master Chief Larry Batten. A tour through the *Texas* sub can take up to a couple of hours and each officer and sailor is capable of explaining the ship and its maze of systems in detail.

On Tuesday Captain John Litherland gave Galveston Mayor Lyda Ann Thomas a personal tour through the ship. Capt. Litherland will hand over command of the *Texas* to Commander James Gray on September 22, in King's Bay Georgia. "Usually morale on a ship under construction isn't as good as a ship deployed," said Capt. Litherland. "But morale on *Texas* has been consistently good."

The *Texas* sailors have been generous with their morale, cooking breakfast for families staying at the Galveston Ronald McDonald House and visiting children in the Shriners Hospital Burn Unit. "There were about fifteen Navy men from the sub who came and spoke with the children," said Clemmie White, a Shriners administrator. "The children love the attention and the sailors left overwhelmed to see these children in good spirits aside from their injuries." Several of the *Texas* crew were bilingual and able to converse with the Spanish-only speaking patients.

Though they have been busy, it hasn't been all work for the *Texas* submariners. Other activities scheduled for the commissioning week include a visit to an area winery, NASA tours, a Texans game, and numerous receptions in their honor.

Typically, a Navy ship pulls into port for liberty, rest, and relaxation. The crew of the *Texas* may have to wait until they go back to sea to get some rest.

When asked which he considered a bigger challenge to *Texas*, a diesel sub with AIP or an advanced nuclear sub, Capt. Litherland said, "With the demise of the Soviet submarine force pretty much complete now, there are not too many threats directly to our submarines. We have the ability to go almost anywhere in the world undetected, we have access that makes us such a potent weapon. When you're in close to the shoreline, as our 21st century missions promise to be, you're closer to folks who... can ruin your day. We think the *Virginia* class submarine is well-equipped for the challenges, particularly working in-shore. We have systems and sensors that enable us to get in there safely, execute a mission undetected, and leave.

This writer asked him to share his personal thoughts as the commanding officer of a nuclear submarine on his role in America's defense. "We in the submarine force don't talk a lot about it," he answered. "We're known as the Silent Service, for good reason. But we have been and continue to see ourselves on the cutting edge, the forefront of the nation's defense."

Silent Hunter II Memoirs

Shawn Storc

Preface

The development of Silent Hunter II could have been considered by any rational person to be a first class Charlie Foxtrot. Despite overwhelming and seemingly insurmountable odds, the team pulled together, burning many barrels of midnight oil, to insure that SHII did not live up to the vaporware expectations that were running rampant in the community. What follows are my own personal recollections of the process of bringing SHII up from crush depth and making it back to port safely, hopefully they will educate, entertain and enlighten you.

How I Got Involved

I was working at SSI as the beta test coordinator and online test dba, a position that offered me contact with every production group in the company. I developed a really good working relationship with the Simulations team due to their large group of external testers working on Flanker 2.0. As a result, when an opening for a Production Assistant came available in that group, I was a natural fit. Projects in the queue at that time were Flanker 2.0, Harpoon4 and Silent Hunter II. I was assigned to SH2, working under Rick Martinez.

SHII History

Before we go any further, it would probably be good to give a quick history lesson. The success of Silent Hunter and Silent Hunter Commander's Edition naturally predicated the creation of a sequel, and what better way to round out a SubSimmer's collection but to take them to the Atlantic Theater? Command Aces of the Deep had done this in the past, but SHII was to encompass all the great features of SHI and CAOD and then some; to deliver a 3D sim that allowed the player to scan the skies and waves with unprecedented freedom, with a 'dynamic campaign' experience, and online play. No short order by any stretch of the imagination. The other major design feature of SHII was the planned interoperability with the companion sim, Destroyer Command, under development by Ultimation, Inc, a first step in realizing the concept of a 'digital battlefield' environment.

The development studio responsible for SHI, Aeon Electronic Entertainment, began work on SHII, and was well into development when I arrived. Unfortunately, some major snags were encountered during development and the end result was that Aeon would not be finishing the game. As the publisher, we were left holding source code and assets for SHII, a project none of us wanted to see die.

After a lot of spreadsheet work, talking, more spreadsheet work, more talking, asset evaluation and yet more talking, an agreement was struck with Ultimation to keep the product from ending up in Davey Jones' locker. What this meant, however, was that Destroyer Command would be put on hold until SHII was finished, since it was the more highly anticipated product of the two titles.

Several complications arose out of this arrangement. First of all was how to technically make SHII happen. Handing source code off to an entirely new dev team may sound easy, but in reality it is no small task. The decision was made to utilize as many of the art assets as possible and rebuild the game using Ultimation's proprietary 'Janus' engine. This would give us the best chance of actually delivering the game, maintaining graphical parity with Destroyer Command, not to mention our best shot at having functional interoperable multiplayer.

If that sounds challenging, just wait, it gets better. Ultimation was also working on Harpoon4 at this time. The introduction of a new and unplanned product into their schedule pushed H4 to 'hold' status until both SHII and DC were out of the way (Harpoon4 was eventually cancelled).

Also taking place at this time were major upheavals in the corporate structure within the publisher. During development, our masthead changed from SSI to Mindscape to Learning Company to Mattel to Gores Technology (under the Broderbund brand) and finally to Ubisoft. If you have ever had to contend with one change of ownership in your place of employment, then you will truly empathize with the uncertainty brought upon by five changes in less than four years.

Due to another consequence of corporate changes, an office move, the entire production team of one, me, ended up 'going native' and moved in to Ultimation's offices. Ubisoft moved everyone from Novato Ca to San Francisco and transferred the majority of testing responsibilities to other Ubi offices either in Montreal or Bucharest Romania. Luckily, I was able to retain a small core team of testers to work onsite. One of the more frustrating things for a tester to hear is a programmer telling them 'it works fine on my machine'. Nothing can replace the benefit of a programmer being able to walk right over and observe a bug being reproduced. We had to pack them in, doubling up in cube spaces at Ultimation, but like the pros they are, nobody complained even though they knew their days were numbered. The test team was invaluable in helping see the project through to completion. Ultimation was located in Petaluma Ca, a short drive from Novato, but a longer drive from SF. Since everyone on the project lived north of SF, Ultimation graciously welcomed us into their office where we set up shop for the duration.

Challenges

If the business situation facing SHII wasn't enough, there was the whole aspect of rebuilding the game on a new engine. A huge amount of research was necessary to attempt to accurately model the systems and behaviors players would be encountering. Artwork and models had to be converted over to work with the Ultimation tool set or created from scratch. Sound effects had to be created, scripts had to be written, voice actors sourced, directed and recorded. The manual had to be written, localization had to take place, functional testing and compatibility had to have both human and capital resources. Then there was that little issue of the 'dynamic campaign'... that went right out the window, we just did not have the time or resources to make it happen. We were under the gun to deliver the game as soon as possible while still maintaining a pride in our work and our responsibility to consumers. Speaking of consumers, relations with the community at times became challenging because of the tough decisions we were making in order to get the project done. No dynamic campaign? No 3-D crew? No wolfpack co-op play? No editor? ...and the list goes on. Thankfully, Ultimation's associate producer, Mark Kundinger, was very active online, serving as the face of development which showed people that the team really cared about the product. I particularly enjoyed watching the tide of attitude change as we were actually able to release screens and videos proving the game really did exist.

Getting back to the difficult situation of cutting features, I won't make excuses, I would have loved to see these things in the game, but they were simply not in the cards. SHII wouldn't see the light of day if we did not adopt a very lean and realistic approach to making the game. This did not mean extra steps weren't taken-- many mornings I would come in and find Ultimation co-owner/uber-programmer Troy Heere already in his office, beckoning me to come see some 'little thing' he had implemented the night before. Troy always amazed us with his ability to figuratively pull rabbits out of his hat with respect to features and functionality. For those of you who love to calculate out your own manual TDC formulas and put them to the test, you have Troy to thank for the fastidious attention to detail in the implementation of the TDC. Other elements like additional mission scripting options, graphical enhancements (the star shells were just so cool) and convoy groups were a small fraction of the extra work done by Troy and the other members of the team to raise the quality bar.

One of the features we could not cut, though it was hotly debated, was interoperable multiplayer with Destroyer Command. If there ever was a 'back of box' feature, this was it. Even if nobody ever played it, it had to be in. Here lies another problem. The original multiplayer middleware purchased to handle the feature was from a company called RTime. In the span of time between purchasing the licenses and the time the feature was actually implemented, RTime had been purchased by Sony and all support ceased to exist. The libraries had been partially integrated so moving over to DirectPlay would mean yet another substantial setback. Thankfully, Ultimation had contracted with a very sharp programmer named Darrel Dearing who was able to actually make the system work. That statement alone does not do justice to the effort involved. Without Darrel, multiplayer would not have happened.

This may sound like we had the problem solved, but nothing is as it seems. You will recall that multiplayer was not available in the shipping version of SHII. That's because we couldn't get it done in time. Secondarily, there was no reason for it-- sub vs. sub combat in WWII was nonexistent and would only be fun to the most masochistic players out there. The rationale was that once DC shipped, we would release a patch that added multiplayer functionality to SHII, and that's exactly what we did.

Solutions

Having smart and dedicated people on the team is a great way to stack the deck in your favor, but no matter how smart or dedicated, nobody can turn back the clock. Throwing more resources at a project is one of the first things teams tend to do, going against common project management logic. The key is in finding quality people instead of quantity. Thankfully Ultimation had people who were not only technically adept, but also fans of history and naval combat, which, as you can imagine, sometimes would get us side-tracked arguing details, tweaking statistics, properties, models and artwork, etc...

On the list of things we needed solutions for were mission and tutorial scripting, writing and proofing the manual, playbalance/playability testing, localization and reality checking. We managed to handle the localization, tutorial and multiplayer mission scripting and manual writing internally (guess who did that, hint, hint) and had to look elsewhere for solutions to getting the rest of the work done.

As I mentioned previously, finding quality not quantity was key, but that was in regard to paid staff. In this situation we had no money to work with, and as we all know, its pretty tough to get something for nothing...

This is where a number of dedicated and die-hard individuals come into play. First on that list is Neal Stevens and a select crew from the subsim boards. Up until this point, I was hesitant to send builds out to anyone, but in speaking with Neal and reading his reviews and articles I felt it was worth a shot. After all, what did I have to loose? Sure enough, upon receiving the build, Neal responded with a great combination of a bug report and playbalance review. Neal offered to hand pick a team of folks from the subsim community to help out and I couldn't refuse, especially since Neal was going to manage the team and send me compiled and edited feedback. We ended up getting valuable fresh eyes on builds as well as a ton of critical feedback that went above and beyond your basic bug reports. Personally, having these folks involved was a breath of fresh air for me, their new perspective and boundless enthusiasm helped insert a dose of reality into a situation which, in the morbid humor of crunch time, had become a death march.

The second group of enthusiasts that pitched in to help was Chris Dean and the crew at Naval Warfare Simulations (NWS). The mission scripting support for SHII was never fully developed and in fact consisted

of an 'editor' which was basically capable of placing units on the map, but not much else. Any further work needed to be conducted in a text editor, which made for a tremendous challenge when creating missions. The NWS guys were very keen on trying their hand with the limited tools we had in hopes of helping take some of this work off of our to do list. We had basically divided the mission building duties for the whole game up between several of us in house, on top of our other responsibilities. Chris and his team gave it a great effort and while all of their work may have not ultimately made it into the game, their feedback on how the system was set up and constant 'what if' questioning helped to enlighten us on ways we could improve the system.

SHII showed me the strength of willpower, professionalism and dedication. It also showed me first hand the damage that these implied 'qualities' can bring if you do not strive for balance. Everyone on the team worked superhuman hours for weeks and months on end to get the job done, no small feat given the level of challenges we had to plot a course through and around.

Fun Stuff and Inside Stories

I believe SHII was shown at a total of 4 E3s. My first experience with the mind numbing sensory overload that was E3 took place in 2001, with the Aeon version of SHII. Being low man on the totem pole at that time, I wasn't scheduled to attend E3. However, Rick Martinez was driving down and offered me a seat in the car. Being young and gullible, I jumped at the chance. If I would have known that I would be pressed into service demoing a very bare bones SHII for the duration of the show and having to sleep on the floor, I would have probably reconsidered. But after all, that's what makes experiences like this memorable.

When I came on to the project, all the travel and research had concluded. Needless to say, I was more than disappointed to have missed the European U-Boat tour and visit with Admiral Topp and the rest of the Kriegsmarine vets at their yearly gathering. We were very lucky to be able to interview these brave gentlemen regarding their experiences during the war. Unfortunately though, due to space constraints on the disk, we were only able to include excerpts from these interviews, predominantly those including Admiral Topp. One of the statements the Admiral made in an interview stuck with the team through the duration of the project and as all good inside jokes go, they make little sense when told in the sobering context of reality. Trust me though, many late nights that turned into early mornings were lightened up by the sound of the good Admiral's thickly accented English calling out 'rabbits!' emanating from someone's cube.

One other tidbit of information which I personally found disappointing was the fact that we had to craft a full 3D sub tour from scratch instead of using an actual VR scene from one of the existing boats. (Owners of Destroyer Command have the benefit of a real VR scene shot at the USS Kidd, located in Baton Rogue, Louisiana. Our first choice was to approach the Museum of Science and Industry in Chicago, IL to see about partnering with them to use their fine VR tour of the U-505 or better yet (for us nerds) to be able to shoot our own. Sadly, neither of these events came to pass, for various reasons, we were unable to come to an equitable agreement with the museum administration. I fondly remember visiting the 505 as a child and strongly suggest it to anyone interested in the subject matter, having that material connection to such a piece of history is a very powerful, and in a small way helps put the experiences of the U-Boat crews into perspective (minus the seasickness, permeating diesel fumes, special U-Boat cocktails and of course, rabbits!)

Mods

The miracle of shipping SHII left us feeling good about actually wrapping a project that was long on the bookies odds sheet as a fool's bet. However the close personal bond we had formed with the project and community left us wishing we could have done more. Earlier, I mentioned the primitive nature of the mission editor and the challenges surrounding the creation of missions. I had little hope for the game to last very long on enthusiasts machines without a proper editor, so I was amazed and encouraged when I began seeing people actually deconstructing the scenario files and trying their hand at hacking the missions.

We ended up releasing all the information about the files to ease the burden of exploratory work and began looking for ways to support the community in these efforts. Neal approached me again with a proposal to create a community funded multiplayer upgrade; all they needed was the game source code. This sounds simple, but source code is the digital dna of any application and companies guard such an asset with fierce resolve. We were in luck though and managed to garner support for the effort from both the publishing side and from Ultimation. The proper papers were drawn up and executed and disks were on their way. Thankfully, due to the efforts of the Subsim team, coder Duane Doutel, and the community that raised $9000 to fund the project, a DirectPlay multiplayer patch dubbed Project Messerwetzer[1] was released, enhancing the stability of the long-sought online battles between destroyers and U-boats.

Acknowledgements

Silent Hunter II was a simple work assignment that turned into a labor of love for everyone involved. I was able to work with many talented, dedicated and inspirational people over the course of my time with SHII and like to think that I'm a bit wiser for the experience. I'd like to take the opportunity to personally thank Carl Norman and Rick Martinez for their guidance, patience and for being good friends to this day. The team at Ultimation (RIP) for their unfaltering resolve to see the project through, especially Dave Bringhurst and Troy Heere. Neal Stevens, the Subsim strike team and NWS proved that you can get something for nothing, good show gentlemen. Last but not least, the entire production and QA team for going the extra mile day after day, all the while knowing that after the project was through, they would be looking for new jobs. I encourage you to go to your bookshelf or game library, pull SHIII off the shelf and open the manual to the credits page, those are the people who made it happen.

Last, I would like to thank you, the enthusiast, for allowing me to travel back in time to re-live both the good and bad times. When I visit the boards these days, and that isn't often any more, it gives me great pleasure and satisfaction to see people continuing to enjoy something I played a small part in making a reality, thank you.

---

[1] Messerwetzer is a German word that means *to sharpen old knives.*

Tales from the Torpedo Room
Don Meadows MMC/SS U.S. Navy (Ret.)

Foreign Ports of Call

I'm from West Virginia. Okay, go ahead let me hear the jokes. If you have none, then here: West Virginia is the home of the toothbrush. Anywhere else and it would have been called the teethbrush...ha-ha-hah. Now, with that out of the way, I continue.

For some reason I was always fascinated by submarines and knew it was my calling. One writer tells of people who are born as 'water people', people who have to be in or near the sea. I must have fallen into that category.

Finally, when old enough I joined the Navy specifically for the Submarine Service. After graduating (as honor man) from sub school I reported to *USS Ray* SSN 653. Four days later I was at sea, throwing up, mess cooking, and wondering what the hell I had gotten myself into.

It was day twenty-one of our mission when I realized that I had never seen the ocean. I was in it but had never seen it. It bothered me and I was determined to get a peek one way or another. My duties in the galley were something from which nightmares are made. Eighteen-hour days feeding men I thought for sure hated me. As a non-qual, air-breathing, water-using heat load, I had yet to earn the respect that submariners demand. I was forbidden to go to control when the boat was at PD, and the thought of a look out the scope was indeed a dream, but see the ocean I would.

Thirty days into the mission, it was announced that the boat would make a port call in Toulon, France. Finally, the day came and we surfaced.

Some communication problem or misunderstanding occurred and we were in early. Only the tugboats came out to meet us. The harbor pilot was an hour away, and he was the only one who spoke English.

To this day, I do not know what happened on the bridge of USS *Ray*. I do know that word was passed down that anyone who spoke even a little French was needed on the bridge.

My chance! My chance to see the ocean. "Here, Chief!" I lied. "Three years of French in High School."

He shrugged. "Well, come on, get your ass to the bridge."

I followed, hoping my smile wasn't too bright. I was going to see the ocean.

Five minutes later, I was atop the sail. The flag fluttered and whipped proudly, and the ocean—my God, this was really the ocean! I couldn't have been happier. All my dreams were now true; I was on top of the world…I then heard the captain's voice.

Now this captain was a cross between Darth Vader and Cujo. He had a habit of slobbering when he yelled, and he yelled a lot.

"*Meadows!* Stop sightseeing! Here, take the bullhorn."

You've all heard of, and some may have had, the phenomenon called 'pucker factor'. Mine was off the scale at that moment.

"We're drifting down to that wharf, and the water is too shallow to use the SPM,"[1] the CO snapped at me. "Tell the tugboats to put the lines over the cleats."

Now, the only French I had ever known in West Virginia was French Fries, and Christina Hill introduced me to a French kiss, but that was it.

I'd seen the ocean and if God were merciful, he would kill me now. I wouldn't have minded.

"*Now* Meadows!" the skipper screamed.

I brought the bullhorn to my lips, and using the best French accent I could conjure, let go with, "*Put ze linez over zee cleats!*"

Those on the tugs looked up at me dumbfounded, but the CO…the CO… I, for the first time in my young life, saw what pure hate looked like.

He was screaming and yelling; he even kicked the bridge hatch as he stuffed me down the bridge access. His booming tirade echoed from all

---

[1] Secondary Propulsion Motor

points of the harbor.

I descended back to the control room and found the helmsman in tears from laughter. The chief of the watch couldn't even talk, and the contact coordinator looked at me from around the periscope as though he'd just seen Elvis.

The next morning the Commanding Officer of the USS Ray SSN 653 held a special Captain's Mast just for me. I was the first enlisted man to be charged with Hazarding a Vessel.

The Real Grapes of Wrath

I turn forty-four this year. I never understood as a child how grownups could care so little or even forget about their own birthdays. Now it hits me. Unfortunately, I now get the big picture.

My medicine chest is slowly filling up. When I was young the worse thing that could happen when I went to the doctor was him sticking a needle in me. Now the worse thing is him sticking his finger…you get the idea. Each year something seems to go wrong. Another ailment reminds you that you aren't a kid anymore.

How many birthdays can you remember? I'm sure you are like me and can remember some of those long gone days. Something happened to make it special. For me it was the day I turned twenty.

Now this story began two months before my birthday. I was serving on the now-gone USS Ray SSN 653. We were, of course, operating as part of the 6th Fleet. The Mediterranean was a great place for young sailors. Too great a place it turned out.

I know I keep getting off subject, but this is supposed to be both entertaining and educational. One of the reasons sailors are kept busy with seemingly mindless and useless tasks is to keep them out of trouble, give them something to do. Because if you don't, here's what happens.

In the many Italian ports we, of course, sampled the wines. 'Sample' is such a gentle word, but it sounds better than 'guzzle'. We grew fond of the grape, very fond. It was at quarters one morning that we received word that we had been ordered home. A few of us plotted a scheme to carry back some of our favorite wines. Ten of us decided the risk would be too great. However, one enterprising member of our little 'Tasting Club', hit upon the idea of us making our own.

So now we have ten guys, all under twenty, serving on a nuclear submarine, at the height of the cold war, plotting to make wine. What could go wrong?

The grapes were easy. We bought those in town. Twenty pounds, if I remember. We stole sugar and yeast from the cooks. All we needed was something in which to make our vintage 1982. While going aft to do some maintenance on the anchor windless I spotted these large plastic bottles just behind the jacking gear. These bottles had a slight yellow hue to them and were about the size and shape of those water bottles you see upside down in coolers. Perfect!

To this day I do not know if the bottles had been used by the nukes or not. If they had been used, then that would sure answer a lot of questions my wife has.

The ship left as scheduled. On the mid-watch, when few people were up, we began. The grapes were crushed and into the bottle they went, then the sugar and the yeast. We topped the bottle off with a rubber glove secured to the neck with a heavy rubber band. That was easy. Now all we had to do was wait.

Of course, we couldn't let the entire boat know of our winery. We knew we had violated about 700 regulations. We had to find a hiding place. If you think there is nowhere to hide anything on a submarine, you are so very wrong.

Under the torpedo tubes there are grates. These grates are behind the tubes and just forward of the mine-handling tables. When a tube is opened there is always water inside. These grates allow the water to fall into the bilges where it can then be pumped into SAN #1. We picked up the grate under the port tube nest. Under the deck was a nice bend in the hydraulic line for the Tube 3 flood and drain valve. Using some line from the TDU we made a nice barrel sling and hung our brew. With the grate in place it was hidden.

Many things which I still cannot discuss happened on our way through the Med. Truth be told, we forgot about our wine. It just sat there aging away in authentic 1982 plastic.

Fridays on submarines are unique. Either in port or at sea always means one thing: 'Field Day'. For those not familiar, Field Day is where the entire crew is awakened and the ship is cleaned for four hours. It was such a Friday, and as it was also my birthday. Why the grape god decided that day was *the day* I'll never know.

As we went about cleaning the torpedo room we received a visit from the executive officer. Our XO at the time was a fearfully mean man. He used to carry a pocket ruler and would measure the length of our hair at random times. He even once had a sailor removed from the boat for being left handed. I kid you not, dear reader. This man was Satan incarnate. When he didn't speak to you, it was a blessing. If you heard his voice, doom was sure to follow.

It was at the exact moment that the gas from the wine's fermentation filled the rubber glove we had secured to the bottle's top. We looked in stunned horror as the XO walked over to the grate, and he saw the same thing we did. A hand sticking out of the bilge, fingers spread and gently waving. For a split second I thought of climbing in tube one, shutting the breech and dying an easier death than I knew would follow.

The XO turned and smiled. "I see someone finally had the guts to clean down there."

What? My mind screamed. I had to say something. "Yes sir, it was getting bad."

Then the anti-Christ leaned over the grate. "Good job down there," he said. Then he turned and went up the forward escape scuttle. We stood in stunned and grateful silence, as if we had been given new life, as if a kidney stone was passed.

After Field Day we retrieved the five-gallon jug. We smelled the top but our noses noted nothing. Had we failed? We quickly strained our wine through a towel into plastic buckets lined with garbage bags. Once filtered, we returned the wine back into the bottle. Some had spilled onto the tile deck of the torpedo room. We failed to notice that it ate the wax, the tile, and probably would have eaten the steel deck.

Now came the moment of truth. The first ever *USS Ray* wine-tasting event was about to start. Since it was my birthday I had the honor of the first sip. Carefully we poured equal amounts in the cups. After a few happy birthdays, I brought the cup to my mouth. It was like grape juice. I was disappointed. I had visions of a great kicking fun, knock-your-socks-off shot of pop skull. The consensus of my shipmates was the same. We pondered where we had gone wrong. We decided that it hadn't aired enough. I had another cup. It was still weak. I shook the bottle and tried another cup. Now it tasted even weaker. Strangely though, I felt my tongue go numb. Then the big mistake, I stood up. My entire body flushed with a heat that I had never felt before. The torpedo room twirled around. The look on my shipmate's faces was suddenly so funny. I laughed at everything. I wanted to sing KISS songs. I managed to feel the boat move at an angle and heard on the 27MC that we were making preparations to go to periscope depth. Oh, how I wanted to go see the pretty lights in control. The last thing I really remember was four of my shipmates each holding one of my limbs and throwing me in my rack.

I woke up thinking it had been only hours to find I had been out for twenty-seven hours. Thank God the corpsman was a kind man and attributed my condition to a sudden case of the flu.

When analyzed by the nukes who used a hydrometer it was determined that our wine was nearly one-and-a-half times stronger than pure grain moonshine. I never knew what happened to the rest of the bottle.

So, friends, if someday we are together and we are offered wine, you will understand if I pass.

Okay, I Was Drunk!

Gibraltar, 1987, *USS Ray* SSN 653. Mission CLASSIFIED (I think). Following a classic, just-like-in-the-movies bar fight in which a local bar tender was tossed none too gently through his own front window, it was decided the good ship *USS Ray* might feel a little safer if she were anchored offshore rather than tied to the pier.

Although I had nothing to do with the above-mentioned activities, (really...trust me...well, maybe a little...all right, all right, I held the bar tender's feet) all who made it back to the boat were guilty. The COB, who hadn't slept in what seemed like a month, was none too happy to be shaken out of his rack with this bit of joyful news.

Once we puttered out from our nice mooring and the anchor let go, the COB decided that we should now suffer. Some were sent to the mess decks to scrub decks, polish bright work, some were sent to the head where they found new uses for their own toothbrushes. COB had a mean streak. I guess I looked the least drunk of them all. How that happened I have no idea. Oh, my punishment was to be a good one.

If you have seen *Das Boot* (and if you are reading this book, the chances are about 100% that you have), the fog comes in like billows of gray cold cotton. Since we were a small ship compared to the tankers, freighters, and warships, it was decided that an extra watch was needed. Guess who drew that?

Up the forward escape trunk I went, COB on my heels bitching up a storm. I was assigned the forward fog watch, on the hull above the torpedo room. I had with me a whistle, and a light jacket I managed to grab. Still feeling the effects of new drink I had discovered (Jagermeister), I asked the COB just how long was I supposed to be on watch. Has anyone ever defined when exactly hell does freeze over? Another blast of professional time-honed profanity marked the COB's leaving me alone in the fog only mere feet from the cold sea.

I was to be on the watch for any ship that might pose a threat to our submarine. It was damn spooky up there! After about an hour of standing and freezing, my legs ached. It was then that I noticed the forward capstan was up. COB never said I couldn't sit down. So I perched myself on the capstan, hoping beyond hope that the COB would have some mercy on me. Another hour came and went.

Suddenly, I looked up and to my horror; I saw the bow of the largest ship I'd ever seen bearing down on us, and me in particular. Everything went into slow motion. I fumbled for what seemed days for the whistle. When I did find it, I blew that whistle till the ball almost shattered. Then I screamed, and I must admit, none too much like a man.

"COLLISION IMMINENT!"

I never thought the collision alarm sounded so good. Below me, I could hear the footsteps of my shipmates, who I hoped could do something and fast. The ship loomed up closer. It seemed like a black wall of steel.

"Where is it?" a voice called from behind me.

I pointed to the ship, knowing that any second the bow of that ship would either cut us in two or at the least knock us into the next world.

Well something did hit me. As a matter of fact, it hit me again and again. It told me what a stupid bastard I was and that this would be my last day on Earth.

A little lesson in the operation of the 637 Class (Short Hull) capstan. If the valve to the hydraulics is not shut off, the capstan turns ever so slightly and oh so quietly.

As I had been sitting there atop the capstan, it had slowly rotated until I was facing, yes, this is true, our *own sail*. Between the Jagermeister, the fog, and the capstan's slow rotation, I had mistaken our very own sail for a ship. I didn't get any sleep that night, and the COB, well, he made me his personal test subject. I discovered areas of that boat that I'd never seen. I leaned to clean things that I thought never needed cleaning. But the most important lesson I learned, and one I still adhere to, is *Jagermiester is evil!*

A Memorable Sight

During my time on *USS Dallas* SSN 700 we made more than our share of SPECOPs to the northern areas. It is inevitable that LANTFLT boats will pull into the Royal Naval Submarine Base, Faslane, Scotland. During such deployment we were in and out of that port more times than not. We kept having problems with this or that, and always required repair.

One morning while pulling in I was in charge of the line handlers on line one (the line up at the front). Our cleats were rolled, and the lines faked out in a neat Flemish. (The Brits have a tradition of seamanship and we were determined to at least equal them). As we stood waiting for the tugs *Dallas* preceded slowly passed Gourock, and further up the Firth of Clyde, past Helensburgh. We passed Holy Loch and made a slight left turn.

Then word came of the JA phones that a submarine would be surfacing off our port bow. What? How could that be? We took up almost the whole waterway. Must be a joke. Oh, were we shocked when the feather of a periscope trailed along the side of our round hull. Then a submarine did indeed surface. A tiny thing when compared to our 688, but let me tell you it sent a shiver down my back. She was a beautiful boat. Simple in line yet powerful and purposeful.

She came to the surface and settled into a perfect trim. Still at speed, she cut across our bow at a range of about 600 yards. There she seemed to stop without the slightest disturbance of water. No sooner had her headway died off than men could be seen running along her deck. As if by magic, an inflatable expanded and slid into the water. The men on the deck seemed as if they were part of the boat as they climbed in. A motor roared to life, and off they went.

The now surfaced *Upholder's* stern squatted as her screw dug into the calm dark waters. In seconds she was at speed. A few seconds later the bow dipped under as a jet of air hissed from her forward ballast tanks. She was handled by the best. Just as her bow slipped under, the after ballast tanks were vented, and she was gone. Only the trail of disturbed water from her attack scope betrayed her, and that too was gone in a wink. All this happened in under a minute. Suddenly we all felt like we were riding on barge instead of a 688. To see that little boat so well handled, it almost brought a tear to the mariner's eye. It is a memory I will carry always.

Air Pressure

Names in this installment have been changed. You'll see why later.

Air is a funny thing. In the outside world you hardly notice it. The same is true on a submarine. Oh, sure, you know what air is and what it does. Twelve psi in the ballast tanks means they are empty. Two thousand psi is great for shooting weapons. Forty psi means you can flush the head. These are just numbers that don't really mean anything. However, when you see air pressure in use, it can and does leave an impression.

It was a rough surface transit from the surfacing point to the pier in Groton. Seas of twenty feet hounded us all the way to the New London Light. The seas were bad, but when you add the freezing wind of a New England winter, it was hell. We had just completed a TRE (Technical Readiness Evaluation.) The entire crew was bone tired and looking forward to some time at home.

As luck would have it, I, of course, had the duty after the ship was tied up and topside rigged. I had it all planned out. Eat some dinner, call the wife, watch a movie, and then hit the rack until time for my topside tour.

We made short order of bringing the ship in. No one wanted to spend any more time in the weather than they had to. Within the hour of the first line going over it was just me and the duty section. The first part of the plan was successful. I ate a nice dinner of spaghetti with a wonderful meat sauce. Afterwards I made my way forward to the goat locker.

I heard the announcement over the 1MC.

"Rig Ship for lady visitors. There will be lady visitors onboard till further notice."

Oh, one bit of information I left out. When it is very cold, a submarine will align its ventilation system in the re-circulate line up. This means the boat is still tight as if it were submerged. This ventilation lineup keeps the warm air in. Air is drawn into the ship via whatever hatch is open. However, this does funny things when the weapons shipping hatch is open.

I had just picked a movie to watch when I felt my ears try to pop. At about the same time I heard the CAMS (Central Air Monitoring System) alarm. Seconds later I heard a commotion in the upper level under the weapons hatch.

Being so ingrained with submarine life I went to the scene, and what a scene it was. There was our cook, (we'll call him "Bill") who was dripping wet. He weighed about 100 pounds. His eyes were wide with terror, as he and the below decks watch wrestled and pushed at something in the hatch.

"What's going on?" I asked, just knowing there were no pleasant answers.

Bill looked at me, his eyes as wide as the plates he served us on. "My wife," he yelled pointing to the hatch. Now let explain Bill's wife. She is a wonderful lady and she and Bill are to this day some of my best friends. However, Mrs. Bill tipped the scale I would say at about 325. On this day, Mrs. Bill had started down the weapons shipping hatch ladder and somehow gotten her…her, well, have you ever heard the term 'butt in a ringer'?

Why I looked up the hatch I will never know. All I saw was stretching spandex. It kinda looked like a huge water balloon that was filling to the bursting point.

Both men had pushed and shoved but the…the lady would not move. The fans were now drawing a vacuum on the ship, wedging her tighter in the hatch. Just so you know, the hatch is 21-inches in diameter. We could hear her somewhat, and she was less than happy. I started feeling pretty bad myself.

I went to control and picked up the growler. For a moment I had thought of calling the duty officer and formally asking permission to 'Surface Ventilate'. However, I thought that explaining Mrs. Bill's butt was stuck in the hatch and the suction on the ship was about to pop the fan room vacuum breakers, might take some time, so I went ahead and did it. I hit the toggle for VH-1 Snorkel Head Valve, and opened VH-2 Induction backup. I actually thought I heard the fans sigh with relief.

Sure that the pressure would equalize and free the...blockage, I went back to the hatch. No, the blockage remained, wedged tight as ever. By now the duty officer had arrived. I quickly briefed him on the status. You could just see the gears in his head turning as he wondered how he was going to explain this to the captain.

All options were now on the table. One idea: grease. Bill went to the galley, retrieved some lard and applied it to the problem. More pushing, and still nothing. Next idea: have someone pull from topside. The unlucky sonar tech couldn't find enough leverage, and that plan went the way of the Gooney.

Last idea before we called for professional help: Salvage air. Salvage air is a 700-psi system designed to keep water out by putting air pressure in. Now we had to be careful with this. Too much air and the local paper's headline for the next day would have said something about a 325-pound lady with a greased butt flying over Naval Submarine Base Groton.

Once again the ship was put in re-circulate and we gently applied the 700-psi till a squishy pop and a moan was heard. Other than a bruise and a great story, Mrs. Bill was none the worse for wear.

So next time you see some number followed by 'psi', think about it. It may just save your butt.

The Subsim community is an example on how to show dedication and interest for your favourite game-genre! The first thing that really struck me when I joined Subsim was the friendly modding community. Everybody was extremely helpful, and I learned a lot (and I'm still learning) from the excellent people at Subsim. Combined with the dedicated Subsim-people I'm proud to call my friends, this one of the best communities in existence on the Internet.

Dan "_Seth_" Steinbakk

The Diary of a U-boat Commander

Sir Stephen King-Hall

With an introduction and explanatory notes by Etienne

INTRODUCTION

"I would ask you a favour," said the German captain, as we sat in the cabin of a U-boat which had just been added to the long line of bedraggled captives which stretched themselves for a mile or more in Harwich Harbour, in November, 1918.

I made no reply; I had just granted him a favour by allowing him to leave the upper deck of the submarine, in order that he might await the motor launch in some sort of privacy; why should he ask for more?

Undeterred by my silence, he continued: "I have a great friend, Lieutenant-zu-See Von Schenk, who brought U.122 over last week; he has lost a diary, quite private, he left it in error; can he have it?"

I deliberated, felt a certain pity, then remembered the Belgian Prince and other things, and so, looking the German in the face, I said:

"I can do nothing."

"Please."

I shook my head, then, to my astonishment, the German placed his head in his hands and wept, his massive frame (for he was a very big man) shook in irregular spasms; it was a most extraordinary spectacle.

It seemed to me absurd that a man who had suffered, without visible emotion, the monstrous humiliation of handing over his command intact, should break down over a trivial incident concerning a diary, and not even his own diary, and yet there was this man crying openly before me.

It rather impressed me, and I felt a curious shyness at being present, as if I had stumbled accidentally into some private recess of his mind. I closed the cabin door, for I heard the voices of my crew approaching.

He wept for some time, perhaps ten minutes, and I wished very much to know of what he was thinking, but I couldn't imagine how it would be possible to find out.

I think that my behaviour in connection with his friend's diary added the last necessary drop of water to the floods of emotion which he had striven, and striven successfully, to hold in check during the agony of handing over the boat, and now the dam had crumbled and broken away.

It struck me that, down in the brilliantly-lit, stuffy little cabin, the result of the war was epitomized. On the table were some instruments I had forbidden him to remove, but which my first lieutenant had discovered in the engineer officer's bag.

On the settee lay a cheap, imitation leather suit-case, containing his spare clothes and a few books. At the table sat Germany in defeat, weeping, but not the tears of repentance, rather the tears of bitter regret for humiliations undergone and ambitions unrealized.

We did not speak again, for I heard the launch come alongside, and, as she bumped against the U-boat, the noise echoed through the hull into the cabin, and aroused him from his sorrows. He wiped his eyes, and, with an attempt at his former hardiness, he followed me on deck and boarded the motor launch.

Next day I visited U.122, and these papers are presented to the public, with such additional remarks as seemed desirable; for some curious reason the author seems to have omitted nearly all dates. This may have been due to the fear that the book, if captured, would be of great value to the British Intelligence Department if the entries were dated. The papers are in the form of two volumes in black leather binding, with a long letter inside the cover of the second volume.

Internal evidence has permitted me to add the dates as regards the years. My thanks are due to K. for assistance in translation.

--ETIENNE.

Torplexed

The Diary of a U-boat Commander

One volume of my war-journal completed, and I must confess it is dull reading.

I could not help smiling as I read my enthusiastic remarks at the outbreak of war, when we visualized battles by the week. What a contrast between our expectations and the actual facts.

Months of monotony, and I haven't even seen an Englishman yet.

Our battle cruisers have had a little amusement with the coast raids at Scarborough and elsewhere, but we battle-fleet fellows have seen nothing, and done nothing. So I have decided to volunteer for the U-boat service, and my name went in last week, though I am told it may be months before I am taken, as there are about 250 lieutenants already on the waiting list.

But sooner or later I suppose something will come of it. I shall have no cause to complain of inactivity in that Service, if I get there.

I am off tonight for a six-days trip, two days of which are to be spent in the train, to the Verdun sector. It has been a great piece of luck. The trip had been arranged by the Military and Naval Inter-communication Department; and two officers from this squadron were to go. There were 130 candidates, so we drew lots; as usual I was lucky and drew one of the two chances.

It should be intensely interesting.

⊕ ⊕ ⊕

At ----

I arrived here last night after a slow and tiresome journey, which was somewhat alleviated by an excellent bottle of French wine which I purchased whilst in the Champagne district.

Long before we reached the vicinity of Verdun it was obvious to the most casual observer that we were heading for a centre of unusual activity. Hospital trains traveling north-east and east were numerous, and twice our train, which was one of the ordinary military trains, was shunted on to a siding to allow troop trains to rumble past.

As we approached Verdun the noise of artillery, which I had heard distantly once or twice during the day, as the casual railway train approached the front, became more intense and grew from a low murmur into a steady noise of a kind of growling description, punctuated at irregular intervals by very deep booms as some especially heavy piece was discharged, or an ammunition dump went up.

The country here is very different from the mud flats of Flanders, as it is hilly and well wooded. The Meuse, in the course of centuries, has cut its way through the rampart of hills which surround Verdun, and we are attacking the place from three directions. On the north we are slowly forcing the French back on either river bank--a very costly proceeding, as each wing must advance an equal amount, or the one that advances is enfiladed from across the river.

We are also slowly creeping forward from the east and north-east in the direction of Douaumont.

I am attached to a 105-cm. battery, a young Major von Markel in command, a most charming fellow. I spent all today in the advanced observing position with a young subaltern called Grabel, also a nice young fellow. I was in position at 6 a.m., and, as apparently is common here, mist hides everything from view until the sun attains a certain strength. Our battery was supporting the attack on the north side of the river, though the battery itself was on the south side, and firing over a hill called L'Homme Mort.

Von Markel told me that the fighting here has not been previously equaled in the war, such is the intensity of the combat and the price each side is paying.

I could see for myself that this was so and the whole atmosphere of the place is pregnant with the supreme importance of this struggle, which may well be the dying convulsions of decadent France. His Imperial Majesty himself has arrived on the scene to witness the final triumph of our arms, and all agree that the end is imminent.

Once we get Verdun, it is the general opinion that this portion of the French front will break completely, carrying with it the adjacent sectors, and the French Armies in the Vosges and Argonne will be committed to a general retreat on converging lines. But, favourable as this would be to us, it is generally considered here that the fall of Verdun will break the moral resistance of the French nation. The feeling is, that infinitely more is involved than the capture of a French town, or even the destruction of a French Army; it is a question of stamina; it is the climax of the world war, the focal point of the colossal struggle between the Latin and the Teuton, and on the battlefields of Verdun the gods will decide the destinies of nations.

When I got to the forward observing position, which was situated among the ruins of a house, a most amazing noise made conversation difficult. The orchestra was in full blast and something approaching 12,000 pieces of all sizes were in action on our side alone, this being the greatest artillery concentration yet effected during the war.

We were situated on one side of a valley which ran up at right angles to the river, whose actual course was hidden by mist, which also obscured the bottom of our valley. The front line was down in this little valley, and as I arrived we lifted our barrage on to the far hill-side to cover an attack which we were delivering at dawn.

Nothing could be seen of the conflict down below, but after half an hour we received orders to bring back our barrage again, and Grabel informed me that the attack had evidently failed. This afternoon I heard that it was indeed so, and that one division (the 58th), which had tried to work along the river bank and outflank the hill, had been caught by a concentration of six batteries of French 75's, which were situated across the river. The unfortunate 58th, forced back from the river-side, had heroically fought their way up the side of the hill, only to encounter our barrage, which, owing to the mist, we thought was well above and ahead of where they would be.

Under this fresh blow the 58th had retired to their trenches at the bottom of the small valley. As the day warmed up the mist disappeared, and, like a theatre curtain, the lifting of this veil revealed the whole scene in its terrible and yet mechanical splendour. I say mechanical, for it all seemed unreal to me. I knew I should not see cavalry charges, guns in the open, and all the old-world panoply of war, but I was not prepared for this barren and shell-torn circle of hills, continually being freshly, and, to an uninformed observer, aimlessly lashed by shell fire. Not a man in sight, though below us the ground was thickly strewn with corpses. Overhead a few aeroplanes circled round amidst balls of white shell bursts.

During the day the slow-circling aeroplanes (which were artillery observing machines) were galvanized into frightful activity by the sudden appearance of a fighting machine on one side or the other; this happened several times; it reminded me of a pike amongst young trout.

After lunch I saw a Spad shot down in flames, it was like Lucifer falling down from high heavens. The whole scene was enframed by a sluggish line of observation balloons. Sometimes groups of these would hastily sink to earth, to rise again when the menace of the aeroplane had passed. These balloons seemed more like phlegmatic spectators at some athletic contest than actual participants in the events.

I wish my pen could convey to paper the varied impressions created within my mind in the course of the past day; but it cannot. I have the consolation that, though I think that I have considerable ability as a writer, yet abler pens than mine have abandoned in despair the task of describing a modern battle. I can but reiterate that the dominant impression that remains is of the mechanical nature of this business of modern war, and yet such an impression is a false one, for as in the past so today, and so in the future, it is the human element which is, has been, and will be the foundation of all things.

Once only in the course of the day did I see men in any numbers, and that was when at 3 p.m. the French were detected massing for a counter-attack on the south side of the river. It was doomed to be still-born. As they left their trenches, distant pigmy figures in horizon blue, apparently plodding slowly across the ground, they were lashed by an intensive barrage and the little figures were obliterated in a series of spouting shell bursts. Five minutes later the barrage ceased, the smoke drifted away and not a man was to be seen. Grabel told me that it had probably cost them 750 casualties. What an amazing and efficient destruction of living organism!

Another most interesting day, though of a different nature.

Today was spent witnessing the arrangements for dealing with the wounded. I spent the morning at an advanced dressing station on the south bank of the river. It was in a cellar, beneath the ruins of a house, about 400 yards from the front line and under heavy shell-fire, as close at hand was the remains of what had been a wood, which was being used as a concentration point for reserves. The cover afforded by this so-called wood was extremely slight, and the troops were concentrating for the innumerable attacks and counter-attacks which were taking place under shell fire. This caused the surgeon in charge of the cellar to describe the wood as our main supply station! I entered the cellar at 8 a.m., taking advantage of a partial lull in the shelling, but a machine-gun bullet viciously flipped into a wooden beam at the entrance as I ducked to go in. I was not sorry to get underground. A sloping path brought me into the cellar, on one side of which sappers were digging away the earth to increase the accommodation. The illumination consisted of candles set in bottles and some electric hand lamps. The centre of the cellar was occupied by two portable operating tables, rarely untenanted during the three hours I spent in this hell. The atmosphere--for there was no ventilation--stank of sweat, blood, and chloroform.

By a powerful effort I countered my natural tendency to vomit, and looked around me. The sides of the cellar were lined with figures on stretchers. Some lay still and silent, others writhed and groaned. At intervals, one of the attendants would call the doctor's attention to one of the still forms. A hasty examination ensued, and the stretcher and its contents were removed. A few minutes later the stretcher--empty--returned. The surgeon explained to me that there was no room for corpses in the cellar; business, he genially remarked, was too brisk at the present crucial stage of the great battle. The first feelings of revulsion having been mastered, I determined to make the most of my opportunities, as I have always felt that the naval officer is at a great disadvantage in war as compared with his military brother, in that he but rarely has a chance of accustoming himself to the unpleasant spectacle of torn flesh and bones.

This morning there was no lack of material, and many of the intestinal wounds were peculiarly revolting, so that at lunch-time, when another convenient lull in the torrent of shell fire enabled me to leave the cellar, I felt thoroughly hardened; in fact I had assisted in a humble degree at one or two operations. I had lunch at the 11th Army Medical Headquarters Mess, and it was a sumptuous meal to which I did full justice. After lunch, whilst waiting to be motored to a field hospital, I happened to see a battalion of Silesian troops about to go up to the front line. It was rather curious feeling that one was looking at men, each in himself a unit of civilization, and yet many of whom were about to die in the interests thereof.

Their faces were an interesting study. Some looked careless and debonair, and seemed to swing past with a touch of recklessness in their stride, others were grave and serious, and seemed almost to plod forward to the dictates of an inevitable fatalism. The field hospital, where we met some very charming nurses, on one of whom I think I created a distinct impression, was not particularly interesting. It was clean, well-organized and radiated the efficiency inseparable from the German Army.

⊕ ⊕ ⊕

Back at Wilhelmshaven--curse it!

Yesterday morning, when about to start on a tour of the ammunition supply arrangements, I received an urgent wire recalling me at once! There was nothing for it but to obey. I was lucky enough to get a passage as far as Mons in an albatross scout which was taking dispatches to that place. From there I managed to bluff a motor car out of the town commandant--a most obliging fellow. This took me to Aachen where I got an express.

The reason for my recall was that Witneisser went sick and Arnheim being away, this has left only two in the operations ciphering department. My arrival has made us three. It is pretty strenuous work and, being of a clerical nature, suits me little. The only consolation is that many of the messages are most interesting. I was looking through the back files the other day and amongst other interesting information I came across the wireless report from the boat that had sunk the Lusitania.

It has always been a mystery to me why we sank her, as I do not believe those things pay.

Arnheim has come back, so I have got out of the ciphering department, to my great delight. I have received official information that my application for U-boats has been received. Meanwhile all there is to do is to sit at this ---- hole and wait.

2nd June, 1916.

I have fought in the greatest sea battle of the ages; it has been a wonderful and terrible experience. All the details of the battle will be history, but I feel that I must place on record my personal experiences.

We have not escaped without marks, and the good old König brought 67 dead and 125 wounded into port as the price of the victory off Skajerack, but of the English there are thousands who slept their last sleep in the wrecked hulls of the battle cruisers which will rust for eternal ages upon the Jutland banks.

Sad as our losses are--and the gallant Lutzow has sunk in sight of home--I am filled with pride. We have met that great armada the British Fleet, we have struck them with a hammer blow and we have returned. I was asleep in my cabin when the news came that Hipper was coming south with the British battle cruisers on his beam. In five minutes we were at our action stations. We made contact with Hipper at 5.30 p.m.,[1] and Beatty turned north with his cruisers and fast battleships and we pursued.

Two of the great ships had been sunk by our battle cruisers, and we had hopes of destroying the remainder, when at 6.55 the mist on the northern horizon was pierced by the formidable line of the British Battle Fleet. Jellicoe had arrived!

Three battle cruisers became involved between the lines, and in an instant one was blown up, and another crawled west in a sinking condition. Sudden and terrible are events in a modern sea-battle. Confronted with the concentrated force of Britain's Battle Fleet we turned to east, and for twenty minutes our High Seas Fleet sustained the unequal contest. It was during this period that we were hit seventeen times by heavy shell, though, in my position in the after torpedo control tower, I only realized one hit had taken place, which was when a shell plunged into the after turret and, blowing the roof off, killed every member of the turret's crew.

From my position, when the smoke and dust had blown away, I looked down into a mass of twisted machinery, amongst which I seemed to detect the charred remains of bodies. At about 7.40 we turned, under cover of our smoke screen, and steered south-west. Our position was not satisfactory, as the last information of the enemy reported them as turning to the southward; consequently they were between us and Heligoland.

At 11 p.m. we received a signal for divisions of battle fleets to steer independently for the Horn Reef swept channel. Ten minutes later we underwent the first of five destroyer attacks.

The British destroyers, searching wide in the night, had located us, and with desperate gallantry pressed home the attack again and again. So close did they come that about 1.30 a.m. we rammed one, passing through her like a knife through a cheese.

It was a wonderful spectacle to see those sinister craft, rushing madly to their destruction down the bright beam of our powerful searchlights. It was an avenue of death for them, but to the credit of their Service it must stand that throughout the long nightmare they did not hesitate. The surrounding darkness seemed to vomit forth flotilla after flotilla of these cavalry of the sea. And they struck us once, a torpedo right forward, which will keep us in dock for a month, but did no vital injury.

---

[1] This is 4.30 G.M.T.--Etienne

When morning dawned, misty and soft, as is its way in June in the Bight, we were to the eastward of the British, and so we came honourably home to Wilhelmshaven, feeling that the young Navy had laid worthy foundations for its tradition to grow upon.

We are to report at Kiel, and shall be six weeks upon the job.

Frankfurt.

Back on seventeen days' leave, and everyone here very anxious to hear details of the battle of Skajerack. It is very pleasant to have something to talk to the women about. Usually the gallant field greys hold the drawing-room floor, with their startling tales from the Western Front, of how they nearly took Verdun, and would have if the British hadn't insisted on being slaughtered on the Somme. It is quite impossible in many ways to tell that there is a war on as far as social life in this place is concerned. There is a shortage of good coffee and that is about all.

Arrived back on board last night. They have made a fine job of us, and we go through the canal to the Schillig Roads early next week. We are to do three weeks' gunnery practices from there, to train the new drafts.

1916 (about August).

At last! Thank Heavens, my application has been granted. Schmitt (the Secretary) told me this morning that a letter has come from the Admiralty to say that I am to present myself for medical examination at the board at Wilhelmshaven tomorrow.

What joy! to strike a blow at last, finished for ever the cursed monotony of inactivity of this High Seas Fleet life. But the U-boat war! Ah! that goes well. We shall bring those stubborn, blood-sucking islanders to their knees by striking at them through their bellies. When I think of London and no food, and Glasgow and no food, then who can say what will happen? Revolt! rebellion in England, and our brave field greys on the west will smash them to atoms in the spring of 1917, and I, Karl Schenk, will have helped directly in this! Great thought--but calm! I am not there yet, there is still this confounded medical board. I almost wish I had not drunk so much last night, not that it makes any difference, but still one must run no risks, for I hear that the medical is terribly strict for the U-boat service. Only the cream is skimmed! Well, tomorrow we shall see.

⊕   ⊕   ⊕

Passed! and with flying colours; it seemed absurdly easy and only took ten minutes, but then my physique is magnificent, thanks to the physical training I have always done. I am now due to get three weeks' leave, and then to Zeebrugge. I have wired to the little mother at Frankfurt.

At Zeebrugge, or rather Bruges.

I spent three weeks at home, all the family are pleased except mother; she has a woman's dread of danger; it is a pleasing characteristic in peace time, but a cloy on pleasure in days of war. To her, with the narrowness of a female's intellect, I really believe I am of more importance than the Fatherland--how absurd. Whilst at Frankfurt I saw a good deal of Rosa; she seems better looking each time I meet her; doubtless she is still developing to full womanhood. Moritz was home from Flanders. He had ten days' leave from Ypres, and, though I have a dislike for him, he certainly was interesting, though why the English cling to those wretched ruins is more than I can understand.

I felt instinctively that in a sense Moritz and I were rivals where Rosa was concerned, though I have never considered her in that light--as yet. One day, perhaps? These women are much the same everywhere, and I could see that having entered the U-boat service made a difference with Rosa, though her logic should have told her that I was no different. But is that right? After all, it is something to have joined this service; the Guards themselves have no better cachet, and it is certainly cheaper.

Here we live in billets and in a commandeered hotel. The life ashore is pleasant enough; the damned Belgians are sometimes sulky, but they know who is master. Bissing (a splendid chap) sees to that. As a matter of fact we have benefited them by our occupation, the shops do a roaring trade at preposterous prices, and shamefully enough the German shopkeepers are most guilty. These pot-bellied merchants don't seem to realize that they exist owing to our exertions. I was much struck with the beautiful orderliness of the small gardens which we have laid out since 1914, and, in fact, wherever one looks there is evidence of the genius of the German race for thorough organization. Yet these Belgians don't seem to appreciate it. I can't understand it.

I find here that social life is very much gayer than at that mad town of Wilhelmshaven. At the High Seas Fleet bases there was the strictness and austerity that some people seem to consider necessary to show that we are at war, though Heaven knows there was precious little war in the High Seas Fleet; perhaps that was why the "blood and iron" régime was in full order ashore. Here, in Bruges, at any rate as far as the submarine officers are concerned, the matter is far different. When the boats are in, one seems to do as one likes, with a perfunctory visit to the ship in the course of the day. Witnitz (the Commodore) favours complete relaxation when in from a trip. In the evenings there are parties, for which there are always ladies, and I find it is necessary to have a "smoking."[2] I went to the best tailor to buy one, and found that I must have one made at the damnable price of 140 marks; the fitter, an oily Jew, had the incredible impertinence to assure me it would be cut on London lines!

I nearly felled him to the ground; can one never get away from England and things English? I'll see his account waits a bit before I settle it.

There are several fellows I know here. Karl Müller, who was 3rd watchkeeper in the Yorck, and Adolf Hilfsbaumer, who was captain of G.176, are the two I know best. They are both doing a few trips as second in commands of the later U.C. boats, which are mine-laying off the English coasts. This is a most dangerous operation, and nearly all the U.C. boats are commanded by reserve officers, of whom there are a good many in the Mess.

Excellent fellows, no doubt, but somewhat uncouth and lacking the finer points of breeding; as far as I can see in the short time I have been here they keep themselves to themselves a good deal. I certainly don't wish to mix with them. Unfortunately, it appears that I am almost bound to be appointed as second in command of one of the U.C. boats, for at least one trip before I go to the periscope school and train for a command of my own. The idea of being bottled up in an elongated cigar and under the command of one of those nautical plough-boys is repellent. However, the Von Schenks have never been too proud to obey in order to learn how to command.

⊕ ⊕ ⊕

---

[2] A dinner jacket.

I have been appointed second in command to U.C.47. Her captain is one Max Alten by name. Beyond the fact that I saw him drunk one night in the Mess I know nothing of him. I reported to him and he seems rather in awe of me. His fears are groundless. I shall make it as easy as possible for him, for it must be as awkward for him as it is unpleasant for me.

To celebrate my proper entry into the U-boat service, I gave a dinner party last night in a private room at "Le Coq d'Or." I asked Karl and Adolf, and told them to bring three girls. My opposite number was a lovely girl called Zoe something or other. I wore my "smoking" for the first time; it is certainly a becoming costume. We drank a good deal of champagne and had a very pleasant little debauch; the girls got very merry, and I kissed Zoe once. She was not very angry. I think she is thoroughly charming, and I have accepted an invitation to take tea at her flat. She is either the wife or the chère amie of a colonel in the Brandenburgers, I could not make out which. Luckily the gallant "Cockchafer" is at the moment on the La Bassée sector, where I was interested to observe that heavy fighting has broken out today. I must console the fair Zoe!

Both Karl and Adolf got rather drunk, Adolf hopelessly so, but I, as usual, was hardly affected. I have a head of iron, provided the liquor is good, and I saw to that point.

We were sailing, or rather going down the canal to Zeebrugge on Friday, but the starting resistance of the port main motor burnt out and we were delayed till Sunday, as they will fit a new one. I must confess the organization for repair work here is admirable, as very little is done by the crews in the U-boats, all work being carried out by the permanent staff, who are quartered at Bruges docks. Taking advantage of the delay I called on Zoe Stein, as I find she is named.

It appears she is not married to Colonel Stein. She told me he was fat and ugly, and laughed a good deal about him. She showed me his photograph, and certainly he is no beauty. However, he must be a man of means, as he has given her a charming flat, beautifully decorated with water-colours which the Colonel salved from the French château in the early days--these army fellows had all the chances.

I bade an affectionate farewell to Zoe, and I trust Stein will be still busily engaged at La Bassée when I return in a fortnight's time! I am greatly obliged to Karl for the introduction, and told him so; he himself is running after a little grass widow whose husband has been missing for some months. I think Karl finds it an expensive game; luckily Zoe seems well supplied with money--the essential ingredient in a joyous life. On Friday night we had an air-raid--a frequent event here, but my first experience in this line. Unpleasant, but a fine spectacle, considerable damage done near the docks and an unexploded bomb fell in a street near our headquarters.

Two machines (British) brought down in flames. I saw the green balls[3] for the first time. A most fascinating sight to see them floating up in waving chains into the vault of heaven; they reminded me of making daisy chains as a child.

At Zeebrugge.

We are alongside the mole in one of the new submarine shelters that has been built. The boat is under a concrete roof over three feet thick, which would defy the heaviest bomb. We have much improved the port since our arrival. The port, so-called, is purely artificial, and actually consists of a long mole with a gentle curve in it, which reaches out to seaward and protects the mouth of the canal. The tides are very strong up and down the coast, and constant dredging is carried out to keep 20 feet of water over the sill at the lock gates.

On arrival last night we went straight into No. 11 shelter, as an air-raid was expected, but nothing happened, so I went up to the "Flandre," which seems to be the best hotel here, full of submarine people, and I heard many interesting stories. There seems no doubt this U-boat war is dangerous work; I find the U.C. boats are beginning to be called the Suicide Club, after the famous English story of that name, which, curiously enough, I saw on the kinematograph at Frankfurt last leave. We Germans are extraordinarily broad-minded; I doubt if the works of German authors are seen on the screens in England or France.

The news from the West is good, the English are hurling themselves to destruction against our steel front. We are now to load up with mines. I must stop writing to superintend this work.

⊕   ⊕   ⊕

---

[3] Known as "Flying-onions."

At sea. Near the South Dogger Light.

We loaded up the ten mines we carry in an hour and five minutes. They were lifted from a railway truck by a big crane and delicately lowered into the mine tubes, of which we have five in the bows. The tubes extend from the upper deck of the ship to her keel, and slope aft to facilitate release. Having completed with fuel at Bruges, we took in a store of provisions and Alten went up to the Commodore's office to get our sailing orders.

We sailed at 6 p.m. and at last I felt I was off. Today, the 22nd, we are just north of the South Dogger, steering north-westerly at 9-1/2 knots. The sea is quite calm and everything is very pleasant. Our mission is to lay a small minefield off Newcastle in the East Coast war channel. I have, of course, never been to sea for any length of time in a U-boat, and it is all very novel. I find the roar of the Diesel engine very relentless, and last night slept badly in a wretched bunk, which was a poor substitute for my lovely quarters in the barracks at Wilhelmshaven. One thing I appreciate, and that is the food; it is really excellent: fresh milk, fresh butter, white bread and many other luxuries.

I have spent most of the day picking up things about the boat. Her general arrangement is as follows:

Starting in the bows, mine tubes occupy the centre of the boat, leaving two narrow passages, one each side. In the port passage is the wireless cabinet and signal flag lockers, with store rooms underneath. In the starboard passage are one or two small pumps and the kitchen.

The next compartment contains four bunks, two each side, these are occupied by Alten, myself, the engineer, and the Navigating Warrant Officer. Proceeding further aft one enters the control room, in which one periscope is situated, and the necessary valves and pumps for diving the boat.

The next compartment is the crew space; ten of the company exist here.

Overhead on each side is the gear for releasing the torpedoes from the external torpedo tubes, of which we carry one each side. I think we borrowed this idea from the Russians.

Then comes the engine-room, an inferno of rattling noises, but excellent engines, I believe. At the after end of the engine-room are the two main switchboards, of whose manner of working I am at present in some ignorance.

The two main sets of electric motors are underneath the boards, in the stern, where we have a third torpedo tube.

I had hardly written the above words when a message came that the captain would like me to come to the bridge. I went up in a leisurely fashion, through the conning tower, which is over the control room, and reported myself. He indicated a low-lying patch of smoke on the horizon far away on the starboard bow. I was obliged to confess that it conveyed nothing to me, when he aroused my intense interest by stating that it was, without doubt, being emitted from a British submarine, who are known to frequent these waters. He was proceeding away from us, and was, even then, six or seven miles away, so an attack was out of the question. The engineer, who had joined us, drew my attention to the thin wisp of almost invisible blue-grey smoke from our own stern. The contrast was certainly striking!

Over dinner I gave it as my opinion that the British boats were pretty useless. Alten would not agree, and stated that, though in certain technical aspects they were in a position of inferiority, yet in personnel and skill in attacking they were fully our equals. He seemed to hold them in considerable respect, and he remarked that, when making a passage, he was more anxious on their account than in any other way. He informed me that, on the last passage he made, he was attacked by a British boat which he never saw, the only indication he received being a torpedo which jumped out of the water almost over his tail. Luckily it was very rough at the time, which made the torpedo run erratically, otherwise they would undoubtedly have been hit.

What appeared to astonish him was the fact that the British boat had been able to make an attack in such weather. We are now charging on one engine, 500 amperes on each half-battery.

We are due back at Zeebrugge at 10 p.m. tonight. We should have been in at dawn today, but we received a wireless from the senior officer, Zeebrugge, to say that mine-laying was suspected, and we were to wait till the "Q.R." channel, from the Blankenberg buoy, had been swept. We lay in the bottom for eight hours, a few miles from the western end of the channel. Our trip was quite successful, but not without certain excitements.

On the night of the 23rd we passed fairly close to a fishing fleet on the Dogger Bank, and saw the lights of several steamers in the distance. As our first business was to lay our mines in the appointed place, we did not worry them. We burnt usual navigation lights, or rather side lights which appear to be usual, except that, by a little fitting which Alten has made himself, the arcs of bearing on which the lights show can be changed at will. His idea is that, should we appear to be approaching a steamer which he wishes to avoid, in many cases, by shining a little more or less red and green light, we can make her think that we are a steamer on such a course that it is her duty by the rules of the road to keep clear of us.

He tells me it has worked on several occasions, and he has also found it useful to have two small auxiliary side lights fitted which are the wrong colours for the sides they are on. It is, of course, only neutral shipping which carry lights nowadays, though Alten says that many British ships are still incredibly careless in the matter of lights. However, to resume my account of what happened. We reached our position at dawn or slightly after, the weather was beautifully calm and the sea like glass. As we were only three miles from the English coast, and close to the mouth of the Tyne, we were extraordinarily lucky to have nothing in sight, if one excepts a long smudge of smoke which trailed across the horizon to the southward.

The land itself was obscured by early morning banks of mist, yet everything was so still that we actually faintly heard the whistle of a train. I could hardly restrain from suggesting to Alten that we should elevate the 10-cm. gun to fifteen degrees and fire a few rounds on to "proud Albion's virgin shores," but I did not do so as I felt fairly certain that he would not approve, and I do not wish to lay myself open to rebuffs from him after his behaviour concerning the smoking incident. I boil with rage at the thought, but again I digress.

The fact that the land was obscured was favourable from the point of view that we were not worried by coast watchers, but unfavourable from the standpoint that we were unable to take bearings of anything and so ascertain our exact position. The importance of this point in submarine mine-laying is obvious, for, owing to our small cargo of eggs, it is quite possible that we may be sent here again, to lay an adjacent field, in which case it is highly desirable to know the exact position of one's previous effort.

We were somewhat assisted in our efforts to locate ourselves by the fact that a seven-fathom patch existed exactly where we had to lay. We picked up the edge of this bank with our sounding machine, and steering north half a mile, laid our mines in latitude--No! on second thoughts I will omit the precise position, for, though I shall take every precaution, there is no saying that through some misfortune this Journal might not get into the wrong hands.

I am very glad I decided to keep these notes, as I shall take much pleasure in reading them when Victory crowns our efforts and the joys of a peaceful life return. I found it a delightful sensation being so close to the enemy coast, in his territorial waters, in fact. For the first time since the Skajerack battle I experienced the personal joys of war, the sensation of intimate and successful contact with the enemy, and the most hated enemy at that.

We had hardly finished laying our eggs when a droning noise was heard. With marvelous celerity we dived, that damned fellow Alten, who, under these circumstances leaves the bridge last, treading on my fingers as he followed me down the conning tower ladder. The engineer endeavoured to sympathize with me, and made some idiotic remark about my being quicker when I had had more practice. I bit his head off. I can't stand this hail-fellow-well-met attitude in these U.C. boats, from any lout dressed in an officer's uniform. They wouldn't be holding commissions if it wasn't for the war, and they should remember that fact. I suppose they think I'm stand-offish. Well, if they had my family tree behind them they would understand.

We dived to sixty feet, and then came up to twenty. Alten looked through the periscope, and then invited me to look. Curiosity impelled me to accept this favour and, putting the focussing lever to "skyscrape" I swept round the sky. At last I saw him; he was a small gas-bag of diminutive size, beneath which was suspended a little car, the most ridiculous little travesty of an airship I have ever seen. He was nosing along at about 800 feet and making about 40 knots.

Suddenly he must have seen the wake of our periscope, for he turned towards us. Simultaneously Alten, from the conning tower (I was using the other periscope in the control room), ordered the boat to sixty feet, and put the helm hard over. We had turned sixteen points,[4] and in about two minutes heard a series of reports right astern of us. It was evident that our ruse had succeeded and that he had overshot the mark.

Inside the boat one felt a slight jar as each bomb went off. We gradually came round to our proper course, and cruised all day submerged at dead slow speed. Every time we lifted our periscope he was still hanging about sufficiently close to make it foolish for us to come to the surface.

Towards noon a group of trawlers, doubtless summoned by wireless, appeared, and proceeded to wander about. These seemed to concern Alten far more than the airship, and he informed me that from their, to me,

---

[4] 180°

aimless movements he deduced they were hunting for us by hydroplanes. Occasionally we lay on the bottom in nineteen fathoms.

By 4 p.m. the atmosphere was becoming rather unpleasant and hot, and gradually we took off more clothes. Curiously enough, I longed for a smoke, but wild horses would not have made me ask Alten for permission.

At 8 p.m. it was sufficiently dark to enable us to rise, which gave me great pleasure, though the first rush of fresh air down the hatch made me vomit after hours of breathing the vitiated muck. On coming to the surface we saw nothing in sight, but a breeze had sprung up which caused spray to break over the bridge as we chugged along at 9 knots. Everyone was in high spirits, as always on the return journey, when the mind turns to the Fatherland and all it holds.

My mind turns to Zoe. I confess it to myself frankly. I hardly realized to what extent this woman had begun to influence me until we received the wireless signal ordering us to delay entering for twelve hours. The receipt of this news, trivial though the delay has been, threw a mantle of gloom over the crew. I participated in the depression and, upon thought, rather wondered that this should be so. Self-analysis on the lines laid down by Schessmanweil[5] revealed to me that the basis of my annoyance is the fact that my next meeting with Zoe is deferred! I feel instinctively that I shall have trouble here, and that I had better haul off a lee shore whilst there is manoeuvring room, and yet--and yet I secretly rejoice that every revolution of the propeller, every clank and rattle of the Diesels brings us closer together.

Alten has just come down from the bridge, and we chatted for some moments; it is evident that he wishes to apologize for his rudeness over the smoking incident. I was in error, I admit it frankly; at the same time I did not know that the battery was on charge, and to dash a match from my hand! I could have shot him where he stood. However, I am not vindictive, and as far as I am concerned the incident is ended.

One thing I find trying in this small boat, and that is that I can find no space in which to do half my Müller exercises, the leg-and-arm-swinging ones. I must see whether I can't invent a set of U-boat exercises!

Good! In two hours we reach the Mole-end light buoy.

⊕   ⊕   ⊕

---

[5] Apparently some German author, of obscure origin, as I cannot find him in any book of reference.--ETIENNE.

Submarine Mess, Bruges.

It is midnight, and as I write in my room at the top of the house the low rumble of the guns from the south-west vibrates faintly through the open window, for it is extraordinarily warm for the time of year, and I have flung back the curtains and risked the light shining.

We spent the night at Zeebrugge and came up to the docks here next day. We shall probably be in for a week, and I am on four days' "extended absence from the boat," which practically means that I can go where I like in the neighbourhood provided I am handy to a telephone.

After a short inward struggle I rang Zoe up on the telephone; fortunately I did not call first. A man's voice answered, and for a moment I was dumbfounded. I guessed at once it was the Colonel, and I had counted so confidently on his being still away at the front. For an instant I felt speechless, an impulse came to me to ring off without further ado, but I restrained myself, and then a fine idea came into my head.

"Who is that?" I said.

"Colonel Stein!" replied the voice, and my fears were confirmed, but my plan of campaign held good.

"I am speaking," I continued, "on behalf of Lieutenant Von Schenk---"

"Ah, yes!" growled the voice, and for an instant a panic seized me, but I resumed:

"He met Madame Stein at dinner some days ago, and she kindly asked him to call; he has asked me to ring up and inquire when it would be convenient, as he would like to meet you, sir, as well. He has been unable to ring up himself, as he was sent away from Bruges on duty early this morning."

I smiled to myself at this little lie and listened.

"Your friend had better call tomorrow then, for I leave tomorrow evening for the Somme front; will you tell him?"

I replied that I would, and left the telephone well satisfied, but cursing the fates that made it advisable to keep clear of No. 10, Kafelle Strasse for thirty-six hours. Needless to say next day I rang up again in order to tell the Colonel that Lieutenant Schenk had apparently been detained, as he was not yet back in Bruges, and how I felt sure that he would be sorry at missing the Colonel, etc., etc., but all this camouflage was unnecessary, as she herself came to the 'phone. I could have kissed the instrument when I told her of my stratagem and heard her silvery laughter in my ear.

"It is arranged that tomorrow, starting at 10.30, we motor for the day to the Forest of Meten, taking our lunch and tea with us--pray Heaven the weather holds."

Tonight in the Mess it is generally considered that U.B.40 has been lost; she is ten days overdue and was operating off Havre, she has made no signal for a fortnight. Such is the price of victory and the cost of war-- death, perhaps, in some terrible form, but bah! away with such thoughts, tomorrow there is love and life and Zoe!

Once more it is night, still the guns rumble on the same old dismal tones, and as it is raining now it must be getting bad up at the front. Except for the rain it might have been last night, but much has happened to me in the meanwhile. Today in the forest by Ruysslede I found that I loved Zoe, loved her as I have never yet loved woman, loved her with my soul and all that is me. The day was gloriously fine when we started, and an hour's run took us to the forest. We left the car at an inn and wandered down one of the glades. I carried the basket and we strolled on and on until we found a suitable place deep in the heart of the forest.

I have the sailor's love for woods, for their depths, their shadows, their mysteries, which are so vivid a contrast to the monotony of the sea, with the everlasting circle of the horizon and the half-bowl of the heavens above. In the forest today, though the leaves had turned to gold and red and brown, the beeches were still well covered, and overhead we were tented with a russet canopy.

I say, at last we found a spot, or rather Zoe, who, with girlish pleasure in the adventure, had run ahead, called to me, and as I write I seem to hear the echoes of "Karl! Karl!" which rang through the wood. When I came up to her she proudly pointed to the place she had found. It was ideal. An outcrop of rock formed a miniature Matterhorn in the forest, and beneath its shelter with the old trees as silent witnesses we sat and joked and laughed, and made twenty attempts to light a fire.

After lunch, a little incident happened which had an enormous effect on me; Zoe asked me whether I would mind if she smoked.

How many women in these days would think of doing that? And yet, had she but known it, I am still sufficiently old-fashioned to appreciate the implied respect for any possible prejudices which was contained in her request.

After lunch, I asked her a question to which I dreaded the answer. I asked her whether, now that the old Colonel had gone to the Somme, whether that meant that she would be leaving Bruges.

She laughed and teasingly said: "Quien sabe, señor," but seeing my real anxiety on this point, she assured me that she was not leaving for the present. The Colonel, she said, had a strange belief that once a man had served on the Flanders Front, and especially on the Ypres salient, he always came back to die there. It appears that the Colonel has done fourteen months' service on the salient alone, and is firmly convinced he will end his career on that great burial ground. As we were talking about the Colonel I longed to ask her how she had met him, and perhaps find out why she lives with him, for I cannot believe she loves him, but I did not dare.

Strangely enough I found that a curious shyness had taken hold of me with regard to Zoe. I said to myself, "Fool! you are alone with her, you long to kiss her; you have kissed her, first at the dinner-party, secondly when you said good-bye at her flat," and yet today it was different.

Then I was kissing a pretty woman, I was on the eve of a dangerous life, and I was simply extracting the animal pleasures whilst I lived. Today it was a case of Zoe, the personality I loved; I still longed to kiss her, but I wanted to have the unquestioned right to kiss her, as much as I wanted the kisses. I wanted to have her for my own, away from the contaminating ownership of the old Colonel, and I determined to get her. I think she noticed the changed attitude on my part, and perhaps she felt herself that a subtle change in our relationship had taken place, and whilst I meditated on these things she fell into a doze at my side.

I was sitting slightly above her, smoking to keep the midges away, and as I looked down on her childish figure a great tenderness for her filled my mind. She is very beautiful and to me desirable above all women; I can see her as she lay there trustfully at my feet. I will describe her, and then, when I get her photograph, I will read this when I am far away on a trip.

She is of average height, for I am just over six feet and she reaches to just above my shoulder. Her hair is gloriously thick and of a deep black colour, and lies low on her forehead. Her complexion is of the purest whiteness beyond compare, which but accentuates the red warmth of the lips which encircle her little mouth. Her figure is slight and her ankles are my delight, but her crowning glories, which I have purposely left till last, are her eyes.

I feel I could lose my soul; I have lost it, if I have one, in the violet depths of those eyes, which were veiled as she slept by the long black eyelashes which curled up delicately as they rested on her cheeks. I have reread this description, and it is oh, so unsatisfying; would I had the pen of a Goethe or a Shakespeare, yet for want of more skill the description shall stand. How I long for her to be mine, and yet, unfortunate that I am, I cannot for certain declare that she loves me. A thousand doubts arise. I torment myself with recollections of her behaviour at the dinner-party, when within two hours of our first meeting she gave me her lips. Yet did I not first roughly kiss her as we danced? I find consolation in the fact that, though she has said nothing, yet her conduct today was different. She was so quiet after tea as we wandered back through the forests with the setting sun striking golden beams aslant the tree trunks.

Before we left I sang to her Tchaikowsky's beautiful song, "To the Forest," and I think she was pleased, for I may say with justice that my voice is of high quality for an amateur, and the song goes well without an accompaniment, whilst the atmosphere and surroundings were ideal. There was only one jarring note in a perfect day; when we returned to the car the chauffeur permitted himself a sardonic grin. Zoe unfortunately saw it and blushed scarlet. I could have struck him on his impudent mouth, but for her sake I judged it advisable to notice nothing. I feel I could go on writing about her all night, but it is nearly 2 a.m. I must get some sleep.

The guns rumble steadily in the south-west, and the sky is lit by their flashes; may the fighting on the Somme be bloody these coming days.

[Probably about ten days later.--Etienne.]

We leave tonight, having had a longer spell than usual. I am in a distracted state of mind. Since our glorious day in the forest I have seen her nearly every afternoon, though twice that swine Alten has kept me in the boat in connection with some replacements of the battery. I have found out that, like me, she is intensely musical. She plays beautifully on the piano, and we had long hours together playing Chopin and Beethoven; we also played some of Moussorgsky's duets, but I love her best when she plays Chopin, the composer pre-eminent of love and passion. She has masses of music, as the Colonel gives her what she likes. We also played a lot of Debussy. At first I demurred at playing a living French composer's works, but she pouted and looked so adorable that all my scruples vanished in an instant, so we closed all the doors and she played it for hours very softly whilst I forgot the war and all its horrors and remembered only that I was with the well-beloved girl.

The Colonel writes from Thiepval, where the British are pouring out their blood like water. He writes very interesting letters, and has had many narrow escapes, but unfortunately he seems to bear a charmed life. His letters are full of details, and I wonder he gets them past the Field Censorship, but I suppose he censors his own. She laughs at them and calls them her Colonel's dispatches; she says he is so accustomed to writing official reports that the poor old man can't write an ordinary letter.

I told her that I thought the way he mentioned regiments and dispositions rather indiscreet, and she agrees, but she says he has asked her to keep them, with a view to forming a collection of letters written from the front whilst the incidents he describes are vivid in his mind. I suppose the old ass knows his own business, and one day the collection may be completed by a telegram "Regretting to announce, etc. etc." The sooner the better.

So the days passed pleasantly enough, and never by a gesture or word of mouth did she show that I was more to her than any other pleasant young man.

I kissed her when I arrived, I kissed her when I left, each day was the same. She would put her arms round my neck and look long and deeply into my eyes, then she would gently kiss my lips. Not an atom of emotion! not a spark from the fires which I feel must be raging beneath that ~~diabolically extraordinary~~[6] amazingly calm exterior.

On ordinary subjects she would chatter vivaciously enough and she can talk in a fascinating manner on every subject I care to bring up, but as soon as I drew the conversation round to a personal line she gradually became more silent and a far-away and distant look came into those wonderful eyes. I have found out nothing about her beyond the fact that she has travelled all over Europe. I don't even know how old she is, but I should guess twenty-six. I tried to find out a few details by means of discreet remarks at the Club and elsewhere. She simply arrived here about a year ago--as a singer, and met the Colonel--beyond that, all is mystery. Everything about her attracts me powerfully, and this mystery adds subtleties to her charms.

This afternoon I went to say good-bye; I told her we were leaving "shortly," and she gently reproved me for disobeying the order which forbids discussion of movements, but I could see she was not greatly displeased. After tea she played to me, music of the modern Russian school--Arensky, Sibelius and Pilsuki; a storm was brewing and we both felt

---

[6] These words are crossed out.--ETIENNE.

sad. She played for an hour or so, and then came and sat by me on a low divan by the fire. We were silent for a long while in the gathering gloom, whilst a thousand thoughts chased each other swiftly through my brain, as I endeavoured to summon up courage to say what I had determined I must say before I left her, perhaps for ever. At last, when only her profile was visible against the glow of the logs, I spoke.

I told her quietly, calmly and almost dispassionately that I had grown to love her and that to me she was life itself. I told her that I had tried not to speak until I could endure no longer. She sat very still as I spoke, and when I had finished there was a long silence and I gently stretched out my hand and stroked her lovely black hair.

At last she rose and with averted face walked across the room, and stood looking at the storm through the big bow windows. I watched her, but did not dare follow. At length she returned to me, and I saw what I had instinctively known the whole time--that she had been crying. I could not think why.

She put her arms round my neck, kissed me on the forehead and murmured, "Poor Karl." I felt crushed; I dared not move for fear of breaking the magic of the moment, yet I longed to know more; I felt overwhelmed by some colossal mystery that seemed to be enveloping me in its folds. Why did she pity me? Why did she weep? Why didn't she answer my avowal? Why didn't she tell me something? Such were some of the problems that perplexed me.

It was thus when the clock chimed seven. I told her that my leave was up at seven o'clock, and that at 7.15 I had to be back on board the boat. She remembered this, and in an instant the past quarter of an hour might never have existed. She was all agitation and nervousness lest I should be late on board--though at the moment I would have cheerfully missed the boat to hear her say she loved me.

I tried to protest, but in vain. With feminine quickness she utilized the incident to avoid a situation she evidently found full of difficulty, and at 7.10, with the memory of a light kiss on my lips and her God-speed in my ears I was in a taxi driving to the docks in a blinding rain-storm--and we sail tonight. For five, six, seven, perhaps ten days at the least, and at the most for ever, I am doomed to be away from her and without news of her. And I don't even know whether she loves me!

I think I can say she cares for me up to a certain point, but I want more.

"Oh Zoe! of the violet eyes,
And hair of blackest night
Thy lips are brightest crimson,
Thy skin is dazzling white.
"Oh! lay your head upon my breast,
And lift your lips to mine;
Then murmur in soft breathings,
Drink deep from what is thine.
"Then let the war rage onward,
Let kingdoms rise and fall;
To each shall be the other,
Their life, their hope, their all."[7]

At sea

We are bound for the same old spot as last time. Alten must have been drinking like a fish lately; his breath smells like a distillery; he is apparently partial to schnapps, which he gets easily in Bruges. I can't help admiring the man, as he is a rigid teetotaller at sea, though he must find the strain well nigh intolerable, judging from the condition he was in when he came on board last night. He was really totally unfit to take charge of the boat, and I virtually took her down the canal, though with sottish obstinacy he insisted on remaining on the bridge.

This morning, though his complexion was a hideous yellow colour, he seems quite all right. I shall play a little trick on him at dinner tonight. I have begun to get to know some of the crew by now; they are a fine lot of youngsters with a seasoning of half a dozen older men. The coxswain, Schmitt by name, is a splendid old petty officer who has been in the U-boat service since 1911. His favourite enjoyment is to spin yarns to the younger members of the crew, who know of his weakness and play up to it. He has a favourite expression which runs thus:

---

[7] I am indebted to Commander C. C. for the above rough translation of Karl's effusion.-- ETIENNE.

"His Majesty the Kaiser said Germany's future lies on the sea; I say Germany's future lies under the sea."

He is inordinately fond of this statement, and the youngsters continually say: "What made you take to U-boat work, Schmitt?" and the invariable reply is as above. When he has been asked the question about half a dozen times in the course of a day, he is liable to become suspicious, and if his questioner is within range Schmitt stares at him for a few seconds in an absent-minded way, then an arm like that of a gorilla shoots out, and the quizzer (Untersucher) receives a resounding box on the ears to the huge delight of his companions. The old man then permits his iron-lipped mouth to relax into a caustic smile, after which he is left in peace for some time.

At the wheel he is an artist, for he seems to divine what the next order is going to be, or if he is steering her on a course he predicts the direction of the next wave even as a skilful chess player works out the moves ahead.

I am rather weary and ought to go to bed, but before I lose the savour I must record the splendid fun I had with Alten at dinner.

We were dining alone, as the navigator was on the bridge, and the engineer was busy with a slight leak in the cooking water service. I have said that, though a heavy drinker by nature, Alten is a strict abstainer at sea. Accordingly I produced a small flask of rum, half-way through dinner, and helped myself to a liberal tot, placing the liquor between us on the table. As the sight met his eyes and the aroma greeted his nostrils, a gleam of joy flashed across his face, to be succeeded by a frown. With an amiable smile I proffered the flask to him, remarking at the same time: "You don't drink at sea, do you?"

In a thick voice he muttered, "No! Yes--no! thank you."

With an air of having noticed nothing, I resumed my meal, but out of the corner of my eye I watched his left hand on the table near the flask. It was most interesting, all the veins stood out like ropes, and his knuckles almost burst through the skin. This went on for about thirty seconds, when he choked out something about needing a breath of fresh air. As he got up his face was brick red, and I almost thought he'd have a fit. Whether by accident or design he pulled the cloth as he got out from between the settee and the table and upset the flask.

He was apparently incapable of apologizing, for he rushed up on deck. A few minutes later the navigating officer came down and asked what was up?

I said: "What do you mean?"

He said: "Well, the Captain came up just now, swearing like a trooper, and told me to get to the devil out of it; it didn't seem advisable to question him, so I got out of it and came down."

I expressed my opinion that the Captain must be feeling sea-sick and was ashamed to say so. I also suggested to the navigator that he should take the Captain a little brandy in case he was not feeling well, but the navigator declared he was going to stay down in the warmth till he was sent for. Alten is a great coarse brute. Fancy allowing a material substance such as alcohol to grip one's mentality. Thank Heaven I have nerves of iron; nothing would affect me!

And now to bed, though I must just read my account of our day in the forest. Darling girl, may I dream of thee.

We laid our mines without trouble at 5 a.m. this morning, though at midnight we had a most unpleasant experience. I was asleep, as it was my morning watch, when I was awakened by the harsh rattle of the diving alarms. The Diesel subsided with a few spasmodic coughs into silence, and as I jumped out of my bunk and groped for my short sea boots, the navigator and helmsman came tumbling down the conning tower, with the navigator shouting, "Take her down," as hard as you like.

The men at the planes had them "hard-to-dive" in an instant. The vents had been opened as the hooters sounded, and Alten, who had jumped into the control room, immediately rang down, "All out on the electric motors."

In thirty seconds from the original alarm we were at an angle of twenty degrees down by the bow, and I had sat down heavily on the battery boards, completely surprised by the sudden tilt of the deck.

It occurred to me that the air was escaping through the vents with a strangely loud noise, but before I could consider the matter further or even inquire the reason for this sudden dive, the noise increased to a terrifying extent, and whilst I prepared myself for the worst it culminated into a roar as of fifty express trains going through a tunnel, mingled with the noise of a high-powered aeroplane engine.

The roar drummed and beat and shook the boat, then died away as suddenly as it came; a moment later there was a severe jar. We had struck the bottom, still maintaining our angle.

I painfully got to my feet and then discovered from the navigator that he had suddenly seen two white patches of foam 800 yards on the starboard bow, which resolved themselves into the bow waves of a destroyer approaching at full speed to ram. We had dived just in time, and her knife-edged bow, driven by 30,000-horse power, had slid through the water a very few feet above our conning tower. Luckily he had not dropped any depth charges. We were not, however, completely free of our troubles, though we had cheated the destroyer.

Examination of the chart, showed the bottom to be mud, and on attempting to move the foremost hydroplanes, the plane motor fuses blew out. This showed that the boat was buried in the mud right up to her foremost planes, which were immovable.

The hydrophone watchkeeper reported that he could still hear fast-running propellers, though probably some distance away, and as this showed that our old enemy was still nosing about we were very anxious not to break surface. We just blew "A."[8] At least we started to blow "A," but Alten wisely decided that, as it was a calm night with a half-moon, the bubbles on the surface might be rather conspicuous, so we stopped the blow and put the pump on. We also flooded "W".[9] This had no effect on her at all.

We then pumped out "Q" and "P," leaving "W" full, and adjusted our trim to give her only three tons negative buoyancy, just enough to keep us on the bottom if she came out of the mud. In this position we went full speed astern on the motors, 1,500 amps on each, and all the crew in the after-compartment. No result. We then pumped the outer diving tanks on the port side to give her a list to starboard. Still she remained fixed.

So at 2 a.m. we decided to risk it and we put a slow blow on all tanks. When she had about fifty tons positive buoyancy she suddenly bucketed up, and, as the motors were running full speed astern at the time, we came up and broke surface stern first. In a few seconds we were trimmed down again, and as a precautionary measure we proceeded for a couple of miles at twenty metres, when, coming up to periscope depth, we surfaced, and finding all clear we proceeded. We were put down by a trawler at dawn, though she never saw us. After half an hour's hanging about she moved off, which was lucky, as she was right on our billet.

---

[8] Probably their foremost internal tank.--ETIENNE.

[9] Presumably their after internal tank.--ETIENNE.

We are now proceeding to a spot somewhat to the eastward of Cape St. Abbs,[10] as we have instructions to do a two-days patrol here and sink shipping.

We ought to start business tomorrow morning.

We should be in tonight, then for my little Zoe!

But I must record what we have done. Already I am getting much pleasure from reading my diary. Strange how it amuses one to see little bits of oneself on paper, and the less garnished and franker the truths the more entertaining it is. The hours here are so long and boring at times that I feel I want to talk intimately with someone. Failing Zoe I turn to my notebooks.

The first steamer we sighted raised high hopes, at least her smoke did, for we saw enough smoke on the horizon to make us think we were to see the Grand Fleet, and we promptly dived. We cruised towards her for about half an hour, and then hung about where we were, as we found that her course would take the ship close to us.

As the situation developed, Alten, who was up in the conning tower at the "A" periscope, gave us a certain amount of information, and we gathered that all this smoke was pouring out of the pipe-stem tunnel of a wretched little English tramp. I found it most irritating, standing in the control room (my action station) and not knowing what was going on.

There is only one good job in a submarine and that is the Captain's. He knows and decides everything. The rest of us are in his hands and take things on trust. I object on principle to my life being held in Alten's hands. It is all very well for the crew, for, to start with, they have no imagination, and to most of them their mental horizon stops at the walls of the boat. Secondly, they have the consolation of mechanical activities; they make and break switches and open and close valves--they work with their hands. An officer has imagination, and only works with his head.

As we attacked the steamer, all one heard was murmurs from Alten, such as: "Raise!" "Lower!" "Take her down to ten metres!" "Half speed!" "Slow!" "Bring her up to five metres!" "Raise!" "Lower!"

I endeavoured to simulate an air of unconcern which I was far from feeling. Not that I was a prey to physical fear; I flatter myself it is so far

---

[10] St. Abbs Head.--ETIENNE

unknown to me, and there was no great danger, but simply that I longed to know what was happening. At length I heard the welcome order:

"Starboard tube. Stand by!"

Which was followed almost immediately by the order: "Fire!"

There was a kind of coughing grunt, and the starboard torpedo proceeded on its errand of destruction. Every ear was strained for the sound of the explosion, but all we were vouchsafed was a torrent of blasphemy from Alten.

The torpedo had jumped clean out of the water a hundred yards short of the steamer, and had then evidently dived under the ship; so I gathered later when Alten had calmed down somewhat. We were about to surface and give her the gun, when luckily Alten took a good sweep round with the skyscraper and discovered one of those wretched little airships about a mile away, coming towards the steamer, which was wailing piteously, on her siren. As the chart showed forty metres we decided to bottom and have lunch.

Over lunch we discussed the misadventure. Alten was loud in his curses of Tanzerman (the torpedo lieutenant at Bruges), from whom he had got the torpedo in guaranteed good condition only forty-eight hours before we sailed. He launched forth into a tirade against the torpedo staff at Bruges, and, warming to his subject, he roundly abused the whole of the depot personnel, whom he stigmatized as a set of hard-drinking, shore-loafing ruffians, who were incapable of realizing that they existed for the benefit of the boats' personnel and "material."

I naturally disagreed, and did so the more readily that I conscientiously disagree with him. I find that there is a tendency on the part of some of these submarine officers, who have been U-boating a long time, to get into narrow grooves. Most reserve officers are not like this, as they have only been in during the war. Alten is an exception; he left the Hamburg-Amerika on two years' half pay in 1912, and was, of course, kept on in 1914. After all, the depot staff are Germans, and as such labour for the Fatherland, and though their work in office and workshop is not so dangerous as ours, on the other hand they have not got the stimulation before their eyes, of glory to be gained. Personally I am of the opinion that the torpedo broke surface because, being fired from the outside tubes, it probably started too shallow, dived deep, recovered shallow and dived deep, broke surface and dived very deep. A sticky motor or sluggish weight would give this effect.

And are these external tubes water-tight? Theoretically, yes, but what of practice? We have been down to forty metres several times during this trip, and not once have we had a chance on the surface of getting at the two external tubes; add to which our depth gear, with the pivots of the weight exposed to water if the tube does flood and then you have rust, corrosion and heaven knows what complications.

I saw a British Mark 11.50 torpedo at the torpedo shop at Bruges the other day, and I was much struck with their deep depth gear, which is of the unrestrained Uhlan type, i.e., weight and valve interdependent. But then the main feature is that the whole gear is contained in a separate water-tight chamber. Our system is certainly a great saving in space, and is much neater in design, whilst I prefer the Uhlan principle of valve conjuncting with weight, but it would be interesting to know whether the British have much trouble with the depth-keeping of their torpedo.

I have written quite a disquisition on depth gears; I must get on with my record of events. After lunch we had a good look round, but the small airship was still hanging about, flying slowly in large circles. We were rather surprised to meet one of these despicable little sausages or "Zeppelin's Spawn," as the navigator calls them, so far from land, and at dark we surfaced and proceeded on one engine on an easterly course, charging the battery right up with the other engine.

Dawn revealed a blank horizon, not a vestige of mast, funnel or smoke in sight. We ambled along in fine though cold weather, and I took advantage of the peacefulness of everything to do a really good series of Müller on the upper deck, stripped to the waist, and allowed the keen air to play its invigorating currents on my torso.

Alten silently watched me from the conning tower, with a sneering expression on his face. The navigator, who is quite a decent youngster, though of no family, was, I could plainly see, struck by my development, and asked to be initiated into the series of exercises. I agreed willingly enough to show them to him. I will confess I wish Zoe could have seen me as I perspired with healthy exercise.

At about 11 a.m. a couple of masts, then two more, then another, appeared above the horizon. The visibility was extreme, so we at once dived and proceeded at full speed, ten metres. We had been going thus for perhaps half an hour when Alten remarked that he would have another look at the convoy. We eased speed, came up to six metres, and Alten proceeded up into the conning tower to use "A" periscope. He had hardly applied his eye to the lens when he sharply ordered the boat to ten metres, accompanying this order with another to the motor room demanding utmost speed (Ausserste Kraft). I went up to the conning tower and found him white with excitement.

"Look!" he exclaimed, pointing to the periscope, entirely forgetful of the fact that we were at ten metres. I looked, and of course saw nothing; furious at the trick I considered he had played on me I turned on him, to be disarmed by his apology.

"Sorry! I forgot! The whole British battle cruiser force is there."

It was now my turn to be excited, and I rushed down to the motor room determined to give her every amp she would take. The port foremost motor was sparking like the devil, rings of cursed sparks shooting round the commutator, but this was no time for ceremony. I relentlessly ordered the field current to be still further reduced. We were actually running with an F.C. of 3.75 amps,[11] for a period, when the sparking assumed the appearance of a ring of fire and, fearing a commutator strip would melt, I ordered an F.C. of five amps.

We thus passed a quarter of an hour full of strain, the tension of which was reflected in the attitude of all the men. Alten had announced his intention of using the stern torpedo tube after his failure in the morning, and the crew of this tube were crouched at their stations like a gun's crew in the last few seconds preparatory to opening fire. The switchboard attendants gripped the regulating rheostatts as if by their personal efforts they could urge the boat on faster. Old Schmitt, at the helm, never lifted his eyes from the compass repeater.

At length: "Slow both!" "Bring her to six metres!" came from the conning tower, to which place I proceeded to hear the news.

Slowly the periscope was raised and I held my breath; a groan came from Alten and he turned away. For a fraction of a second I was almost pleased at his obvious pain, then, sick with disappointment, I took his place. Yes! it was all over. There they were, and with hungry eyes and depressed heart I saw five great battle cruisers, of which I recognized the Tiger with her three great funnels, the Princess Royal, Lion and two others, zigzagging along at 25 knots, at a distance of 12,000 metres, across our bow.

They were surrounded by a numerous screen of destroyers and light cruisers, the former at that range through the periscope appearing as black smudges. It is not often one is permitted such a spectacle in modern war, and I could not tear myself away from the sight of those great brutes, whom I had fought when in the Derflingger at Dogger Bank and again when in the König at Jutland. So near and yet so far, and as they rapidly

---

[11] The lower the field current the faster the motor goes. 3.75 is almost incredibly low for a motor of this type--at least according to British practice.--ETIENNE.

drew away so did all the visions of an Iron Cross. As soon as they were out of sight, we surfaced in order to report what we had seen to Zeebrugge and Heligoland.

Everything seemed against us. I had gone on the bridge with the navigator; Alten, with a face as black as hell, had gone to the wardroom. About ten minutes elapsed when I heard a fearful altercation going on below. I stepped down to find the young wireless operator trembling in front of Alten, who was overwhelming him with a flood of abuse. As I reached the wardroom, Alten shook his fist in the man's face and bellowed:

"Make the d---- thing work, I tell you."

"Impossible, Captain, the main condenser---" the man began.

Purple with rage, Alten seized a heavy pair of parallel rulers, and before I could check him hurled them full in the operator's face. Bleeding copiously, the youth fell to the deck in a stunned condition. It was then, for the first time, that I noticed a half-empty bottle of spirits on the table, which colossal quantity he must have consumed in about a quarter of an hour. Turning to me, this semi-madman pointed to the wireless operator with his foot and growled, "Have him removed."

This I did, and then, lowering the periscope, I ordered the boat to fifteen metres. We proceeded at this depth until 8 p.m., when I was informed that the Captain was in his bunk and wished to see me.

I discovered him with his face to the ship's side, and upon my reporting myself he ordered me, firstly to throw that blasted bottle overboard (an unnecessary proceeding, as it was empty), and secondly to surface and shape course for Zeebrugge.

At midnight he relieved me, apparently perfectly normal.

The wireless operator has been laid up all day and has a nasty cut on the head. The navigator, a great scandal-monger, has heard from the engineer that Alten was speaking to him alone this morning, and the engineer believes that Alten has given him five hundred marks to say he fell down a hatch.

Hooray! Blankenberg buoy has just been reported in sight! Soon I shall see my Zoe!

With what high hopes did I write the last few lines a few hours ago, and how they were dashed to the ground, for on going into the Mess at Bruges I found amongst my letters a note from her, which was terrible in its brevity. She simply said:

"DEAR KARL,

"I am going away for some days, and as I shall be traveling it is no good giving you an address. To our next meeting!

"ZOE."

How horribly vague; not an indication of her destination, her object, or the probable length of her absence. Of course I rushed round to the flat, but found the place shut up. The porter told me she had gone away with her maid. He couldn't say when she'd be back--if at all! I gave him ten marks, and he said she might be away a fortnight. If I'd given him twenty he'd have said a week; he obviously didn't know. I feel I could do anything tonight; any mad, evil thing would appeal to me.

There is a most fearful uproar coming from the guest-room, where a large and rowdy party are entertaining the chorus of a travelling revue company. I saw them when they arrived, horribly common-looking women, with legs like mine tubes.

Another day and still no news; I don't know how I shall stick it. She might have had the softness of heart to write to me. She knows my address.

This evening a letter from the little mother, who asks whether I can find time to go to Frankfurt when I have leave; at the end of the letter she mentions that Rosa has joined the Women's Voluntary Auxiliary Corps of Army Nurses. I suppose she thought she'd like her photograph taken in some fancy uniform as "Rosa Freinland, one of our Frankfurt beauties, now on war work!" Holding the patient's hand is about the only work she intends doing.

Women as a class are the same the world over. We are well supplied with English papers in the Mess here; they come regularly from Amsterdam, and in their pages I see, just as in ours, pictures of the Countess this and the Lord that, photographed in becoming attitudes doing war work. It seems agricultural pursuits are the fashion in England at present--wait till our U-boat war gets its knife well into their fat guts, it will be more than fashionable to work in the fields then.

The British Empire is undeniably a great creation, or rather not so much a creation as a thing arrived at accidentally, but it lacks solidarity. It sprawls, a confused mass of races and creeds, around the world. Its very immensity lays it open to attack, it has a dozen Achilles heels from Ireland to Egypt and South Africa to India.

I met a man only yesterday who was recently at the propaganda department of the Foreign Office, and without going into details he gave me a very good idea of the good work that is going on in Britain's canker spots. Ireland is considered particularly promising to those in the know. Now for an agitated night! To think that a girl should disturb me so!

⊕ ⊕ ⊕

Two days have passed, or, rather, dragged their interminable lengths away, for there is still not a vestige of news. I have been twice to the flat with no result, except to receive a piece of impertinence from the porter the last time I was there.

No news.

⊕ ⊕ ⊕

Still no news, and we sail in forty-eight hours.

At sea, off the Isle of Wight

It is some days since I turned for solace and enjoyment, amidst the discomforts of this life, to my pen and notebook.

What strange tricks fate plays with us, and how lucky it is that one cannot foresee the future. Here I am in U.39-but I must start at the beginning. My last entry was the depressing one of still no news. Well, I have had news, but it was like a drop of water in the mouth of a parched-up man. Another agonizing twenty-four hours passed, and I was sitting in my room about ten o'clock, trying to resign myself to the idea that the next night I should be starting out for my third trip without news of her, when the telephone bell rang. I lifted the receiver and to my amazed joy heard a voice that I could have recognized in a thousand. It was Zoe!

I was quite incapable of any remark, and my confusion was further increased when, after a few "Hello's," which I idiotically repeated, her clear, level tones said: "Is that you, Karl? How are you?" How was I? What a question to ask! I wanted to tell her that I was bubbling with joy, that a thousand-kilogramme load had been lifted from my chest, that my blood was coursing through my veins, that I, usually so cool, was trembling with excitement, that I could have kissed the mouthpiece of the humble instrument that linked us together. Yet I was quite incapable of answering her simple question! I can't imagine what I expected her to say, for upon reflection her remark was a very ordinary one, and indeed under the circumstances quite natural, but, as I say, in actual fact I was tongue-tied.

I suppose I must have said something, for I next remember her saying: "Well, you might ask how I am;" and to my horror I realized that she thought I was being rude! My abject apologies were cut short by her tantalizing laugh, and I understood that the adorable one was teasing me. When at length I made myself believe that I really was talking to this most elusive and delightful woman I wasted no time in suggesting that, late though it was, I might be permitted to go round and see her. She would not permit this, as she said it would create grave scandal, and the Colonel might hear about it upon his return. I pleaded hard and urged my departure in twenty-four hours.

She was firm and reproved me for discussing movements over the telephone. She was right; I was a fool to do so; but Zoe destroys all my caution. However, she said that I might lunch with her next day, and that she had some new music to play to me. I ventured to ask where she had been, but this question was plainly unpleasing to my lady, so I dropped the subject. I blew her a goodnight kiss over the telephone, to which I think I caught an answer, and then she rang off.

Ten minutes had not elapsed, when a messenger entered and informed me that I was wanted at the Commodore's office at once. A strange feeling of uneasiness and that of impending misfortune overcame me. I felt like a naughty school-boy about to interview the headmaster. I followed the messenger into the Commodore's office, and found myself alone with the great man. He was seated at a huge roll-top desk, which was the only article of furniture in a room which was to all intents and purposes papered with large scale charts of the east and south coasts of England and of the Channel and North Sea.

The Commodore was sealing an envelope as I came in; he looked up and saw me, then, without taking any further notice of me, he resumed his business with the envelope. I felt that I was in the presence of a personality, and I was, for "Old Man Max" is one of the ten men who count in the Naval Administration. He had a reading lamp on his desk, and I remember noticing that the light shining through its green shade imparted a yellow parchment-like effect to the top of his old bald head. With dainty care he finished sealing the envelope, then, picking up a telephone transmitter, he snapped "Admiralty!" In about a minute he was connected, and to my astonishment I realized that he was talking to the duty captain of the operations department in Berlin.

His words chilled my heart, for he said: "Commodore speaking! U.39 sails at 2 a.m. for operation F.Q.H.--Repeat." His words were apparently repeated to his satisfaction, for while I was vainly endeavouring to convince myself that I was unconnected with the sailing of U.39, he banged the receiver into place (Old Man Max does everything in bangs) and snapped at me.

"You Lieutenant Von Schenk?"

I admitted I was, and then heard this disgusting news.

"Kranz, 1st Lieutenant U.39, reported suddenly ill, Zeebrugge, poisoning--you relieve him. Ship sails in one hour forty minutes from now--my car leaves here in forty minutes and takes you to Zeebrugge. Here are operation orders--inform Von Weissman he acknowledges receipt direct to me on 'phone. That's all."

He handed me the envelope and I suppose I walked outside--at least I found myself in the corridor turning the confounded envelope round and round. For one mad moment I felt like rushing in and saying: "But, sir, you don't understand I'm lunching with Zoe tomorrow!"

Then the mental picture which this idea conjured up made me shake with suppressed laughter and I remembered that war was war and that I had only thirty-five minutes in which to collect such gear as I had handy--most of my sea things being in U.C.47--and say goodbye to Zoe.

I ran to my room and made the corridors echo with shouts for my faithful Adolf. The excellent man was soon on the scene, and whilst he stuffed underclothing, towels and other necessary gear into a bag he had purloined from someone's room, I rang up Zoe. I wasted ten minutes getting through, but at last I heard a deliciously sleepy voice murmur, "Who's that?"

I told her, and added that I was off; to my secret joy, an intensely disappointed and long-drawn "Oooh!" came over the wire. So she does care a bit, I thought. Mad ideas of pretending to be suddenly ill crossed my mind--anything to gain twenty-four hours--but the Fatherland is above all such considerations, and after some pleasant talk and many wishes of good luck from the darling girl, with a heavy heart I bade her good-night.

The Old Man's car, which is a sixty horse-power Benz, was waiting at the Mess entrance, and once clear of the sentries we raced down the flat, well-metalled road to Zeebrugge in a very short time. The guard at Bruges barrier had 'phoned us through to the Zeebrugge fortified zone, and we were admitted without delay. In three-quarters of an hour from my interview with old Max I was scrambling across a row of U-boats to reach my new ship, U.39.

I went down the after hatch, reported myself to Von Weissman and delivered his orders to him, of which he acknowledged receipt direct to the Commodore according to instructions. Von Weissman is a very different stamp of man to Alten; of medium height, he has sandy-coloured hair, steel-grey eyes and a protruding jaw. He is what he looks, a fine North Prussian, and is, of course, of excellent family, as the Weissmans have been settled in Grinetz for a long period. He struck me as being about thirty years of age, and on his heart he wore the Cross of the second class. I have heard of him before as being well in the running towards an ordre pour le mérite.

An interesting chart is hanging in the wardroom, on which is marked the last resting-place of every ship he has sunk. He puts a coloured dot, the tint of which varies with the tonnage, black up to 2,000, blue from 2,000-5,000, brown 5,000-8,000, green 8,000-11,000, and a red spot with the ship's name for anything over 11,000. He has got about 120,000 tons at present. He opposes the Arnauld de la Perrière school of thought, which pins faith on the gun, and Weissman has done nearly all his work with the good old torpedo.

Altogether, undoubtedly a man to serve with.

The U.39 was in that buzzing and semi-active condition which to a trained eye is a sure indication that the ship is about to sail. Punctually at five minutes to 2 a.m. Weissman went to the bridge, and at 2 a.m. the wires were slipped and we started on a ten days' trip. As the dim lights on the mole disappeared and the ceaseless fountain of star-shells, mingling with the flashing of guns, rose inland on our port beam my mind travelled overland to the flat at Bruges, and I wondered whether Zoe was lying awake listening to the ceaseless rumble of the Flanders cannon. We went on at full speed, as it was our intention to pass the Dover Straits before dawn.

Though our intelligence bureau issues the most alarming reports as to the frightfulness of the defences here I was agreeably surprised at the ease with which we passed. Von Weissman, to whom I had hinted that we might find the passage tricky, rather laughed at my suggestion, and described to me his method, which, at all events, has the merit of simplicity.

He always goes through with the tide, so as to take as short a time as possible, and he always decides on a course and steers it as closely as possible, keeping to the surface unless he sights anything, and diving as soon as anything shows up. Even if he dives he goes on as fast as possible on his course, irrespective of whether he is being bombed or not.

I must say it worked very well last night. We shaped a course to pass five miles west of Gris Nez, and when that light, which for some reason the French had commodiously lit that night, was abeam, we sighted a black object, probably a trawler or destroyer, about half a dozen miles away right ahead. Weissman immediately dived and, without deviating a degree from his course, held on at three-quarters speed on the motors. Some time later the hydrophone watchkeeper reported the sound of propellers in his listeners, and that he judged them to be close at hand, so I imagine we passed very nearly directly underneath whatever it was.

After an hour's submerging we rose, and found dawn breaking over a leaden and choppy sea. Nothing being in sight, we continued on the surface for an hour, charging batteries with the starboard engine (500 amps on each), but at 9 a.m., the clouds lying low and an aerial patrol being frequent hereabouts, we dived and cruised steadily down channel at slow speed, keeping periscope depth.

Several times in the course of the forenoon we sighted small destroyers and convoy craft[12] in the distance, all steering westerly. They were probably returning from escorting troopships over to France last night. In every case we went to sixty feet long before they could have seen our "stick."[13] Weissman is evidently as cautious in this matter as he is hardy in others; the more I see of him the more I like him; he is a man of breeding, and it is of value to serve in this boat.

As I write we are on the surface about ten miles east of the Isle of Wight, still steering down channel. Tonight at midnight we report our position to Zeebrugge, up till now we have maintained wireless silence for fear of the British and French directional stations picking up our signals and fixing our position. After supper this evening Von Weissman explained to

---

[12] Probably "P" boats.--ETIENNE.

[13] Periscope.--ETIENNE.

me the general plan of our operations for the next eight days. Our cruising billet is about 150 miles south-west of the Scillys, at the focal point where trade for Liverpool and Bristol and the up-channel trade diverges. Von Weissman says that this is a plum billet and we should do well.

I feel this is going to be better than those piffling little mine-laying trips, and though we shall be away ten days, it will qualify me for four days' leave in Belgium.

There was nearly an awkward moment last night, or, rather, there was an awkward moment, and nearly an awkward accident. I relieved the navigator at midnight (the pilot is an unassuming individual called Siegel) and took on the middle watch. It was blowing about force 4 from the south-west, and a nasty short, lumpy sea was running which caught us just on the port bow. About once every ten seconds she missed her step with the waves and, dipping her nose into it, shovelled up tons of water, which, as the bow lifted, raced aft and, breaking against the gun, flung itself in clouds of spray against the bridge. In a very few minutes every exposed portion of me was streaming with water.

At about 2 a.m. I had turned my back to the sea for a moment, and my thoughts were for an instant in Bruges, when, on facing forward once again I saw a sight which effectually brought me back to earth. This was the spectacle of two black shapes, evidently steamers, one on either bow, distant, I should estimate, 600 or 700 metres. I had to make a quick decision, and I decided that to fire a torpedo in that sea with any hope of a hit, especially with the boat on surface, was useless; furthermore, that at any moment either of the steamers might sight us from their high bridge and turn and ram.

These thoughts were the work of an instant, and I at once rang the diving bell, and, pushing the look-out before me, in five seconds I was in the conning tower and had the hatch down. I at once proceeded down into the boat, and the first thing that struck my eye was the diving gauge with the needle practically stationary at two metres.

The boat was not going down properly! and for an instant I was rudely shaken, until a cool voice from the wardroom remarked, "Helm hard a-port," an order that was instantly obeyed, and as she began to turn the moving needle on the depth gauge began its journey round the dial. It was the Captain who had spoken. As soon as he heard the diving alarm he was out of his bunk, and a glance at the gauge he has fitted in the wardroom told him we were not sinking rapidly. In an instant he had put his finger on the trouble, which was that we were almost head on to the sea, with the result that he had given the order as stated above, which, bringing us beam on to the sea, had caused her to dive with ease. He is efficiency itself!

As I explained to him what had happened, the noise of propellers at varying distances from us overhead led him to state his belief that we had run into a convoy homeward bound to Southampton from the Atlantic. He approved of my actions in every particular, save only in my omission to bring the boat away from the sea as I began to dive. This morning we are beginning to get the full force of what is evidently going to be a south-westerly gale of some violence. The seas are getting larger as we debouch into the Atlantic. This looks bad for business.

At the moment we are practically hove to on the surface, with the port engine just jogging to keep her head on to sea and the starboard ticking round to give her a long, slow charge of 200 amps. The wind is force 7-8 and a very big sea is running which makes it entirely impossible to open the conning tower hatch; the engine is getting its air through the special mushroom ventilator, which is apparently not designed to supply both the boat's requirements and those of the engine; the whole ventilator gets covered with sea every now and then, during which period until the baffle drains get the water away no air can get in, so the engine has a good suck at the air in the boat, the result of all this being a slight vacuum in the boat. It is a very unpleasant sensation, and made me very sick. This is really a form of sickness due to the rarefied air.

I had a great surprise when I looked at the barograph this morning as the needle had gone right off the paper at the bottom, and at first glance I thought we had struck a tropical depression of the first magnitude, which, flouting all the laws of meteorology, had somehow found its way to the English Channel; but the engineer explained to me that, as I have already stated, the low atmospheric pressure in the boat was due to the conning-tower hatch being shut down.

I have discovered that Von Weissman is a martyr to sea-sickness--all day he has been lying down as white as a sheet and subsisting on milk tablets and sips of brandy; yet such is the man's inflexibility of will that he forces himself to make a tour of inspection right round the boat every six hours, night and day. It is this will to conquer which has made Germans unconquerable, though "Come the four corners of the world in arms" against us, as the great poet says.

We are, of course, keeping watch from inside the conning tower; it is, at all events, dry, but as to seeing anything one might as well be looking out through a small glass window from inside a breakwater! To bed till 4 a.m.

A most unprofitable day. I grudge every day away from Zoe on which we do nothing. This morning about noon the gale blew itself out, but a heavy confused sea continued to run.

At 2 p.m. we saw a most tantalizing spectacle. A big tank steamer, fully 600 feet long and of probably 17,000 tons burthen hove in sight, escorted by two destroyers. To attack with the gun was impossible, as we could only keep the conning tower open when stern to sea, and in any case the two destroyers prevented any surface work. We tried to get in for an attack, but we had not seen her in time, and the best we could do was to get within 3,000 yards, at which range it would have been absurd to have wasted a torpedo, the chances of hitting being 100 to 1 against, even if the torpedo had run properly in the sea that was on.

I had a good look at her through the foremost periscope in between the waves, and it maddened me to see all that oil, doubtless from Tampico for the Grand Fleet, going safely by. The destroyers were having a bad time of it, crashing into the sea like porpoises, their funnels white with salt, and their bridges enveloped in sheets of water and spray. They little thought that, barely a mile away, amidst the tumbling, crested waves a German eye was watching them!

There is no doubt these damned British have pluck, for it was the last sort of weather in which one would have expected to find destroyers at sea, and yet I suppose they do this throughout the winter. After all, one would expect them to be tough fellows--they are of Teutonic stock--though by their bearing one might imagine that the Creator made an Englishman and then Adam.

Let's hope we get some decent weather tomorrow. I have just been refreshing my memory by reading of what I wrote in the book, concerning the day in the forest with the adorable girl. There is an exquisite pleasure in transporting the mind into such memories of the past when the body is in such surroundings as the present, if only I could will myself to dream of her!

⊕   ⊕   ⊕

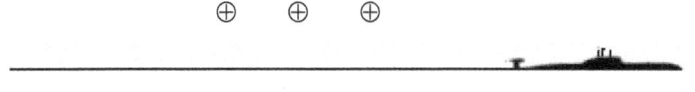

A fine day in every sense of the word. The weather has been and remains excellent, and I have been present at my first sinking. It was absurdly commonplace. At 10 a.m. this morning a column of smoke crept upwards from the southern horizon.

Von Weissman steered towards it on the surface until two masts and the top of a funnel appeared. We dived and proceeded slowly under water on a southerly course.

Half an hour passed and Von Weissman brought the boat up to periscope depth and had a look. He called to me to come and see, an invitation I accepted with alacrity. With natural excitement I looked through the periscope and there she was, unconsciously ambling to her doom like a fat sheep. She was a steamer (British) of about 4,000 tons, slugging home at a steady ten knots, but she was destined to come to her last mooring place ahead of schedule time!

We dipped our periscope and I went forward to the tubes. Five minutes elapsed and the order instrument bell rang, the pointer flicking to "Stand by." I personally removed the firing gear safety pin and put the repeat to "Ready." A breathless pause, then a slight shake and destruction was on its way, whilst I realized by the angle of the boat that Weissman was taking us down a few metres. That shows his coolness, he didn't even trouble to watch his shot.

Anxiously I watch the second hand of my stop watch. Weissman had told me the range would be about 500 metres--30 seconds--31--32--33--has he missed?--34--35--3--A dull rumble comes through the water and the whole boat shakes. Hurra! we have hit, and the order "Surface" comes along the voice pipe.

The cheerful voice of the blower is heard, evacuating the tanks; I run to the conning tower and closely follow Weissman up the ladder. At last I am on the bridge. There she is! What a sight! I feel that I shall never forget what she looked like, though, if all goes well, I shall see many another fine ship go to her grave.

But she was my first; I felt the same sensation when, as a boy, I shot my first roe-deer in the Black Forest, one instant a living thing beautiful to perfection, the next my rifle spoke and a bleeding carcase lay beneath the fine trees. So with this ship. I am a sailor, and to every sailor every ship that floats has, as it were, a soul, a personality, an entity; to carry the analogy further, a merchant craft is like some fat beast of utility, an ox, a cow, or a sheep, whilst a warship is a lion if she is a battleship, a leopard if she is a light cruiser, etc.; in all cases worthy game.

But War has little use for sentimentality! and in my usual wandering manner I see that I have meandered from the point and quite forgotten what she did look like.

What I saw was this:

I saw that the steamer had been hit forward on the starboard side. The upper portion of the stem piece was almost down to the water level, her foremost hold was obviously filling rapidly. Her stern was high out of water, the red ensign of England flapping impotently on the ensign staff. Her propeller, which was still slowly revolving, thrashed the water, and this heightened the impression that I was watching the struggles of a dying animal. The propeller was revolving in spasmodic jerks, due, I imagine, to the fast failing steam only forcing the cranks over their dead centres with an effort.

A boat was being lowered with haste from the two davits abreast the funnel on one side, but when she was full of men and, due to the angle of the ship, well down by the bow, someone inboard let go the foremost fall or else it broke, for the bows of the boat fell downwards and half a dozen figures were projected in grotesque attitudes into the sea. For a few seconds the boat swung backwards and forwards, like a pendulum.

When she came to rest, hanging vertically downwards from the stern, I noticed that a few men were still clinging like flies to her thwarts. Truly, anything is better than the Atlantic in winter. Meanwhile the ship had ceased to sink as far as outward signs went. I mentioned this to Von Weissman, who was at my side with a slight smile on his face, amused doubtless at the eagerness with which I watched every detail of this, to me, novel tragedy. He answered me that I need not worry, that she was being supported by an air lock somewhere forward, that the water was slowly creeping into her and her boilers would probably soon go.

This remarkable man was absolutely correct.

There was an interval of about five minutes, during which another boat, evidently successfully lowered from the other side, came round her stern, picked up one or two men from the water and also collected the survivors in the hanging boat; then the steamer suddenly sank another two feet, there was a dull rumbling, as of heavy machinery falling from a height, a muffled report, a cloud of steam and smoke, a sucking noise and then a pool in the water, in the middle of which odd bits of wood and other buoyant debris kept on bobbing up. Nothing else!

No! I am wrong, there were two other things: a U-boat, representing the might of Germany, and a whaler with perhaps twenty men in it, representing the plight of England!

As she went I felt hushed and solemn, it was an impressive moment; a slight chuckle came from imperturbable Weissman; he had seen too many go to think much of it, and he gave an order for the helm to be put over, so that we might approach the whaler. They were horribly overcrowded, and were engaged in trying to sort themselves into some sort of order. We passed by them at 50 yards and Weissman, seizing his megaphone, shouted in English: "Goodbye! steer west for America!" A cold horror gripped my heart. It was an awful moment. I dare not write the thoughts that entered my head.

I turned away my head and faced aft, that he should not see my face; looking back I saw the whaler rocking dangerously in our wash, and then a commotion took place in her stern, from which a huge bearded man arose and, shaking his fist in our direction, shouted something or other before his companions pulled him down.

Von Weissman heard and his lips narrowed in. I held my breath in suspense, but he evidently decided against what he had been about to do, for with the order, "Course north! ten knots," he went below.

I remained on deck watching the rapidly receding whaler through my glasses until she was a mere speck--alone on the ocean, 150 miles from land, Then the navigator came up, and with strangely mixed feelings of exultant joy and depressing sorrow I went below. Von Weissman was in the wardroom. I watched him unobserved. He was humming a tune to himself and had just completed putting a green dot on the chart. This done he lay back on the settee and closed his eyes--strange, insoluble man!

For long hours I could not forget that whaler; I see it now as I write. I suppose I shall get used to it all. What would Zoe say? The most wonderful thing about man is that he can stand the strain of his own invention of modern war!

I am rather tired tonight, but must just jot down briefly what has taken place today, as there is never any time in the daylight hours.

Soon after dawn, at about 8 a.m., we sighted a fair-sized steamer of about 3,000 tons, which we sunk, but I cannot say what she looked like, or whether anyone escaped, as we never came to the surface at all, Von Weissman sighting smoke on the western horizon just as he hit her. We accordingly steered in that direction. However, I think she went almost at once as Von Weissman put a dot (black) on the chart as we made towards number 3.

I very much wanted to know whether there were any survivors, but I did not like to ask him at the time and he has been in such an infernal temper ever since that I haven't had a suitable opportunity.

The cause of his rage was as follows: Steamer number 3 turned out to be a fine fat chap (of the Clan Line, Von Weissman said, when we first sighted her). We moved in to attack and fired our port bow tube. I waited in vain by the tubes for the expected explosion--nothing happened, but after a couple of minutes a snarl came down the voice pipe: "Surface, GUN ACTION STATIONS!"

I ran aft, and found the Captain white with rage. "Missed ahead!" he said, with intense feeling, "I'll have to use that confounded gun."

In about three minutes the Captain and myself were on the bridge and the crew were at their stations round the gun. For the first time I saw the ship; she was stern on and apparently painted with black and white stripes. As I examined her through glasses--she was distant about 3,000 yards--I saw a flash aboard her and a few seconds later a projectile moaned overhead and fell about 6,000 yards over. So she is armed, thought I, and she has actually opened fire on us first. The effect of this unexpected retort on the part of the Englishman was to throw Weissman into a paroxysm of rage.

"Why don't you fire? What the devil are you waiting for?" etc., etc., were some of the remarks he flung at the gun crew.

I did not consider it advisable to mention to him that they were probably waiting his order to fire, and also his orders for range and deflection, as I had imagined that, here as everywhere else, an officer controls the gun-fire. Apparently in this boat it is not so, as Weissman takes so little interest in his gun that he affects to be, or else actually is, ignorant of the elements of gun control. At any rate, under the lash of his tongue, the gun's crew soon got into action, the gun-layer taking charge. Our first shot was short, very considerably so, as was also the second. Meanwhile the steamer had been keeping up a very creditably controlled rate of fire, straddling us twice, but missing for deflection, as was natural considering that we were bows on to her.

I felt thoroughly in my element listening to the significant wail of the enemy's shell, punctuated by the ear-splitting report of our own gun. Weissman, gripping the rail with both hands, and to my surprise ducking when one went overhead, watched the target with a fixed expression, but made no attempt to control our gun-fire, which was far from creditable, as is inevitable when it is left to the mercy of the inferior intellect of a seaman.

However, at the tenth or eleventh round we hit her in the upper works, as was shown by a bright red and yellow flash near her funnel. This did not check her firing or speed in the least, in fact she seemed to be gaining on us. She also began to zigzag slightly and throw smoke bombs overboard, which were not so effective from her point of view as I had thought they would be.

Matters were thus for some minutes. We had just hit her aft for the second time, though the shooting was so disgustingly bad that I was about to ask whether I might do the duties of control officer, when there was a blinding flash and the air seemed filled with moaning fragments. When I had recovered from my relief from finding that I was personally uninjured, I observed that two of the gun's crew were wounded and one was lying, either killed or seriously wounded, on the casing. We had been hit in the casing, well forward, and, as was subsequently proved when we dived, little material damage was caused to the boat.

This enemy success caused a temporary cessation of fire. The two wounded men were cautiously making their way aft to the conning tower, and I called for a couple of stokers to come up and carry away the third, when Von Weissman suddenly gave the order to dive. The gun's crew at once made a rush for the conning tower, and were down the hatch in a trice, one of the wounded men fainting at the bottom.

I was unaware as to the reason of this order to dive, and thought that perhaps the Captain had sighted a periscope. As I was turning to precede him down the conning tower hatch I distinctly saw the man lying by the gun lift his hand. I felt I could not leave him there, and instinctively cried, "He is still alive!" But Von Weissman, who was urging the crew to hurry down the hatch, pressed the diving alarm as soon as the last sailor was half in the hatch.

I knew that this meant that the boat would be under in 30 to 40 seconds, so I had no alternative but to get down the hatch as quickly as possible. I did so with reluctance, and I was followed by Von Weissman, who joined me in the upper conning tower.

I forced myself not to look out of the conning tower scuttles during the few seconds that elapsed as the casing slowly went under, until at last nothing but waving green water showed at each little window. I feared that, if I had looked, I would have seen a wounded man, stung into activity by the cold touch of the Atlantic. Perhaps Von Weissman read my thoughts, or else he remembered my remark concerning the man, for he turned to me and in level tones said:

"Have you any doubt that he was dead?"

I hesitated a moment, and he continued:

"By my direction you have no doubt. He was!"

How brutal war is, and what a perfect exponent of the art the Captain proves himself to be! To me a life is a life, a particle of the thing divine; to him a life is a unit, and a half-maimed and probably dying seaman is as nothing in the scales when the safety of a U-boat is at stake. The seamen are numbered in their tens of thousands, the U-boats in their tens. The steamer had hit us once, luckily only in the casing, a second hit might well have punctured the pressure hull, and our fate in these waters would have been certain. Therefore, having summed these things up and balanced them in his mind, he dived and the sailor died.

Once below water Von Weissman seemed more his imperturbable self, and unless I am mistaken he is never really happy on the surface, at least when in action. He is a true water mole.

⊕   ⊕   ⊕

A day full of interest, though once again I have had to force myself to absorb the horrors of War. I imagine that I am now going through the experiences of a new arrival on the Western Front, who feels a desire to shudder at the sight of every corpse.

At 10 a.m. this morning we sighted the topsails of a sailing boat to the southwest. Closing her on the surface, we approached to within about 6,000 metres, when suddenly Von Weissman ordered "Gun Action Stations."

The gun crew came tumbling up, but not quick enough to suit him, for as they were mustering at the gun he gave the order to dive, only, however, taking her down to periscope depth before instantly ordering surface and then "Gun Action Stations" again. This time we opened fire on the ship, which was a Norwegian barque and, being in the barred zone, liable to destruction.

Von Weissman had announced overnight that at the first opportunity he would give "that ----- gun's crew a bellyful of practice," and he certainly did. As soon as the first shot was fired, she backed her topsails, and when our fourth shot struck her, somewhere near the foot of the foremast, her crew could be seen hastily abandoning their ship.

This action on their part had no influence with Von Weissman, who had taken personal charge of the helm, and, with the engines running at three-quarter speed, he was zigzagging about, to make it harder for the gun's crew. Every now and then he flung a gibe at the crew, such as suggesting that they should go back to the High Seas Fleet and learn how to shoot.

The sailing ship was soon on fire, for, considering the circumstances, the shooting was very fair, though had I been controlling it I could have confidently guaranteed better results. When she was blazing nicely fore and aft, Von Weissman ordered the practice to cease, and sent the crew below. He then ordered course south, speed ten knots, and I took over the watch. An hour and a half later, when the navigator gave me a spell, a black cloud on the northern horizon marked the funeral pyre of another of our victims. When I went below, the Captain had just finished playing with his precious old chart.

We received a message at 2 a.m. last night from Heligoland to return forthwith; it is now 2 a.m. and we are approaching the redoubtable Dover Barrage. We had no trouble coming up channel today, which seems singularly empty, at any rate in mid-channel, where we were.

We got back

...about three hours ago, and as I was appointed temporary to the boat, Von Weissman kindly allowed me to leave her and come up to Bruges as soon as we got into the shelters at Zeebrugge. I got up here just, in time for a late dinner. Hunger satisfied, I retired to my room and, needless to say, at once rang up my darling Zoe. By the mercy of providence she was in, but imagine my sensations when I heard that that accursed swine of a Colonel was also back from the front, and expected in at the flat at any moment, being then, she thought, engaged in his after dinner drinking bouts at the cavalry officers' club. I could only groan.

A laugh at the other end stung me to furious rage, appeased in an instant by her soothing tones as she told me that I should be glad to hear that he was only up from the Somme on a four-days leave, and was returning next morning by the 8 a.m. troop train. Glad! I could have danced for joy. I breathed again. As the Colonel was expected back at any moment she thought it advisable to terminate the conversation, which was done with obvious reluctance on her part, or so I flatter myself. He goes tomorrow, so far so good, but what of the intervening period?

Could any more refined torture be imagined than that I, who love her as I love my own soul, should have to sit here, whilst scarcely a mile away, probably at this very moment as I write, that gross brute is privileged to kiss her, to look at her, to--oh! it's unbearable. When I think of that hog, for though I've never seen him, I've seen his photograph, and I know instinctively that he is gross, fresh, as she says, from a drinking bout, should at this moment be permitted to raise his pigs' eyes and look into those glorious wells of violet light; when I think that his is the privilege to see those masses of black hair fall in uncontrolled splendour, then I understand to the full the deep pleasures of murder. I would give anything to destroy this man, and could shake the Englishman by the hand who fires the delivering bullet!

Steady! Steady! What do I write? No! I mean it, every word of it. Yet of all the mysteries, and to me Zoe is a mass of them, surely the strangest of all is contained in the question: Why does she live with him? She doesn't love him, she's practically told me so. In fact, I know she doesn't. Let me reason it out by logic. She lives with him, whether voluntarily or involuntarily. Suppose it be voluntarily, then her reasons must be (a) Love; (b) Fascination; (c) Some secret reason. If she is living with him involuntarily it must be: (d) He has a hold on her; (e) For financial reasons.

I strike out at once (a) and (e), for in the case of (e) she knows well that I would provide for her, and (a) I refuse to admit, (b) is hardly credible--I eliminate that. I am left with (c) and (d) which might be the same thing. But what hold can he have on her; she can't have a past, she is too young and sweet for that. I must find out about this before I go to sea again.

Three days ago, I was racking my brains for the solution of a problem, and, as I see from what I wrote, I was somewhat outside myself. In the interval things have taken an amazing turn. I am still bewildered--but I must put it all down from the beginning.

The Colonel left as she said he would, and I went round to lunch with her. We had a delightful tête-à-tête, and after lunch she played the piano. I was feeling in splendid voice and she accompanied me to perfection in Tchaikowsky's "To the Forest," always a favourite of mine. As the last chords died away, Zoe jumped up from the piano and, with eyes dancing with excitement, placed her hands on my shoulders and exclaimed:

"Karl! I have an idea! I shall make a prisoner of you for two or three days."

I laughed heartily and almost told her that she had already made me a prisoner for life, only I can never get those sort of remarks out quick enough. But when she said, "No! I am not joking, I mean it," I felt there was more meaning in her sentence than I had at first thought. I begged to be enlightened, and she then unfolded her scheme.

She told me for the first time, that in a forest not far from Bruges she had a little summer-house, to which she used to retreat for week-ends in the hot weather when the Colonel was away. He knew nothing of this country house (she was very insistent on that point), so I imagined she paid for it out of her dress allowance or in some other way. The idea that had just struck her was that she had a sudden fancy to go and spend two days there, and I was to go with her.

I was ready to go to Africa with her if my leave permitted, and it so happened that I was due for four days' overseas leave (limited to Belgian territory) so that this fitted in very well, and I told her so. She was delighted, then, with one of those quick intuitions which women are so clever at, she read the half-formed thought in my mind, and said: "You mustn't think it's not going to be conventional; old Babette will be with us to chaperon me." Old Babette is an aged female whom she calls her maid. I think she is jealous of me.

I agreed at once that of course I quite understood it was to be highly conventional, etc., though I smiled to myself as I visualized my mother's shocked face and uplifted hands had she heard my Zoe's ideas on the conventions. I was trying to fathom what was at the bottom of it all when she remarked: "Of course, as my prisoner you will have to obey all my orders." I replied that this was certainly so. "And one of the first things," she continued, "that happens to a prisoner when he goes through the enemy lines is that he is blindfolded, and in the same way I shan't let you know where you are going." Seeing a doubtful look in my eyes as I endeavoured to keep pace with the underlying idea, if any, of this truly feminine fancy, she suddenly came up to me and, lifting her eyes to mine, murmured: "Don't you trust me?" In a moment my passion flared up, and rained hot kisses on her face as she struggled to release herself from my arms.

When I left that night after dinner, and, walking on air, returned to the Mess, it was arranged that I should be at her flat with my suit-case at 6 p.m.

the next evening, prepared, to use her own words, "to disappear with me for 48 hours."

She had told me of an address in Bruges which she said would forward on any telegram if I was recalled, and I had to be satisfied with that, for I may as well say here that I never discovered where I went to, and I don't know to this moment in what part of Belgium I spent the last two nights. I tried to find out at first, but as she obviously attached some importance to keeping the locality of her woodland retreat a secret, probably to circumvent the Colonel, I soon gave up trying to get the secret from her, and contented myself with taking things as they came.

To go on with my account of what happened--which was really so remarkable that I propose writing it out in detail to the best of my memory--at 6 p.m. next day I was naturally at her flat feeling very much as if I was on the threshold of an adventure. Zoe was excited and the flat was in a turmoil, as apparently she had only just begun to pack her dressing-case. Soon after six we went down and got into a large Mercédès car which I had noticed standing outside when I arrived. We were soon on our way, and left Bruges by the Eastern barrier; we showed our passes and proceeded into the darkened country-side. We had been running for about a mile when she remarked, "Prisoners will now be blindfolded!" and, to my astonishment, slipped a little black silk bag over my head.

I was so startled I didn't know whether to be angry, or to laugh, or what to do. Eventually I did nothing, and, entering into the spirit of the game, declared that even a wretched prisoner had the right not to be stifled, whereupon she lifted the lower portion of the bag and uncovered my mouth. Shortly afterwards I was electrified to feel a pair of soft lips meet mine, a sensation which was repeated at frequent intervals, and, as I whispered in her ear, under these conditions I was prepared to be taken prisoner into the jaws of hell.

This pleasant journey had lasted for about three-quarters of an hour when my mask was removed and I was informed that I was "inside the enemy lines!" Through the windows of the car I could dimly see that an apparently endless mass of fir trees were rushing past on each side. This state of affairs continued for a kilometre or so, when we branched to the right and soon entered a large clearing in the forest, at one side of which stood the house. Babette, Zoe and myself entered the building, and the car disappeared, presumably back to Bruges.

The house, built of logs, was of two stories; on the ground floor were two living rooms, and the domains of Babette, who amongst her other accomplishments turned out to be not only a most capable valet, but a first-class cook. On the second story there were two large rooms. The whole house was furnished after the manner of a hunting lodge, with stags' heads on the walls, and skins on the floors. In the drawing-room there was a piano and a few etchings of the wild boar by Schaffein.

I dressed for dinner in my "smoking," though under ordinary circumstances I should have considered this rather formal, but I was glad I did, for she appeared in full evening tenue. She wore a violet gown, and across her forehead a black satin bandeau with a Z in diamonds upon it. It must have cost two thousand marks, and I wondered with a dull kind of jealousy whether the Colonel had given it to her.

I cannot remember of what we talked during dinner. We have a hundred subjects in common, and we look at so many aspects of the world through the same pair of eyes; I only know that when I have been talking to her for a period--there is no exact measurement of time for me when I am with her--I leave her presence feeling "completed." I feel that a sort of gap within my being has been filled, that a spiritual hunger has been satisfied, that I have got something which I wanted, but for which I could not have formulated the desire in words. I had resolved that on this first night I would bring matters between us to a head and end this delicious but intolerable uncertainty as to how we stood; yet, when old Babette had served us with coffee in the drawing-room, as I call the second living-room, and we were alone together, I could not bring up the subject. Partly because I think she prevented me so doing by that skilful shepherding of the conversation into other paths with an artfulness with which God endows all women, and also partly because I could not screw myself up to the pitch. I could not, or rather would not, put my fate to the touch. I had a presentiment that in reaching for the summit I might fall from the slope. Alas! how true was this foreboding in some senses--but I will keep all things in their right order.

Let it only be recorded that when she kissed me good-night (with the tenderness of a mother) and left me to smoke a final cigar I had said nothing, and I could only wonder at the strange fate that had placed me practically alone with a girl whom I had grown to love with a deep emotion, and who appeared to love me, yet often behaved as if I was her brother.

The next day we were like two children. The snow was deep on the ground, and the fir trees stood like thousands of sentinels in grey uniform round the clearing. Once during the afternoon, as with Zoe's assistance I was furiously chopping wood for the fire, a droning noise made me look up, and thousands of metres overhead a small squadron of aeroplanes, evidently bound for the Western Front, sailed slowly across the sky. I thought how awkward it would be for them if they experienced an engine failure whilst over the forest, though they were up so high that I imagine they could have glided ten kilometres, and as I think (but I am not certain, and I have pledged myself not to try and find out) we were in the Forest of Montellan, which is barely fifteen kilometres broad, I suppose they could have fallen clear of the trees. As a matter of fact I imagine they would have used our clearing--I'm glad they didn't.

That night after dinner she played to me, first Beethoven and then Chopin. I can see her as I write; she had just finished the 14th Prelude and, resting her chin on her hand, she smiled mysteriously at me.

The hour had come, and, driven by strong impulses, I spoke. I told her that I loved her as I had never thought that a man could love a woman; I told her that I longed to shield her and protect her, and above all things to remove her from the clutches of that bestial Colonel, and as I bent over her and felt my senses swim in the subtleties of her perfume, I begged her passionately to say the word that would give me the right to fight the world on her behalf.

When I had finished she was silent for a long while, and I can remember distinctly that I wondered whether she could hear the thump! thump! thump! of my heart, which to my agitated mind seemed to beat with the strength of a hammer. At length she spoke; two words came slowly from her lips:

"I cannot."

I was not discouraged. I could see, I could feel, that a tremendous struggle was raging, the outward signs of which were concealed by her averted head. At length I asked her point-blank whether she loved me. Her silence gave me my answer, and I took her unresisting body into my arms and kissed her to distraction. Oh! these kisses, how bitter they seem to me now, and yet how I long to hold her once again. For, freeing herself from my embrace and speaking almost mechanically, she said:

"Karl! I must tell you. I cannot marry you."

I pleaded, I prayed, I argued, I demanded. It was in vain; I always came up against the immovable "I cannot."

And then I crashed over the precipice towards whose edge I had been blindly going. I had said for the hundredth time, "But you know you love me," when with a sob she abandoned all reserve, and, flinging her arms round my neck, implored me to take her. Then, as I caught my breath, she quickly said, as if frightened that she had gone too far, "But I cannot marry you."

I looked down into those beautiful eyes, and for the first time I understood. For perhaps ten seconds I battled for my soul and the purity of our love; then, tearing my sight from those eyes which would lure an archangel to destruction, I was once more master of my body. As my resolution grew, I hated her for doing this thing that had wrecked in an instant the hopes of months, the ideals on which I had begun to build afresh my life.

She felt the change, and left me. As she went out by the door she gave me one last look, a look in which love struggled with shame, a look which no man has ever earned the right to receive from any woman.

But I was as a statue of marble, dazed by this calamity.

As the door closed upon her, I started forward--it was too late. Had she waited another instant--but there, I write of what has happened and not what might have been.

I did not sleep that night, until the dawn began to separate each fir tree from the black mass of the forest. Twice in the night, with shame I confess it, I opened my door and looked down the little passage-way; and twice I closed the door and threw myself upon my bed in an agony of torment. It was ten o'clock when a knock at the door aroused me, and the sunlight through the window-pane was tracing patterns on the floor.

There was a note on the breakfast table, but before I opened it I knew that, save for Babette, I was alone in the house. The note was brief, unaddressed and unsigned. I have it here before me; I have meant to tear it up but I cannot. It is a weakness to keep it, but I have lost so much in the last few days, that I will not grudge myself some small relic of what has been. The note says:

> I am leaving for Bruges at half-past eight, when the car was ordered to fetch us back. I go alone. Babette will give you breakfast. The car will return for you at eleven o'clock. I rely on your honour in that you will not observe where you have been. Come to me when you want me--till then, farewell.

It was as she said, and I honourably acceded to her request. This afternoon just before lunch I arrived in Bruges, and since tea-time I have tried to write down what has happened since I left the day before yesterday. Oh! how could she do it, how can it be possible that she is a woman like that? I could have sworn that she was not like this--and yet how can I account for her life with the Colonel? There must be some reason, but in

Heaven's name, what?

Meanwhile I am to go to her when I want her! And that will be when I can give her my name. But oh! Zoe, I want you now, so badly, oh! so badly!

⊕ ⊕ ⊕

I saw her once today in the gardens, walking by herself.

⊕ ⊕ ⊕

I have told Max's secretary that I want to get to sea; to be here in Bruges and not to see her is more than I can bear. I sail at dawn tomorrow. Shall I see her? No, it is best not.

A frightful noise over the New Year celebrations tonight. Champagne flowing like water in the Mess. I feel the year 1917 opens badly for me. Weissman also went to sea again for a short trip in the Channel, and has not reported for five days. Perhaps he has despised the Dover Barrage once too often. If this is so, it is a great loss to the service: he was a man of iron resolution in underwater attack.

I feel I ought to despise Zoe, but I can't. I love her too much; after all, am I not perhaps encasing myself in the robe of a Pharisee? She offered me all she had, save only the one thing I asked, without which I will take nothing. I cannot reconcile her behaviour with her character; why can't she trust me? why can't she be frank with me? I will not believe she is that sort.

I feel I cannot go out again without a sign--I may not return, and I will not leave her, perhaps forever, with this bitterness between us.

At sea

In U.C.47 again. Alten as surly as ever. I decided finally to write to Zoe, but found it difficult to know what to say. Eventually I said more than I had intended. I told her frankly that I experienced a shock, but that I had not meant to seem so cold, and that what I had done had been done for both our sakes. I told her that I still loved her, and I implored her once more to leave the Colonel and come to me as my wife. Already I long to know what message awaits me on my return.

This will not be for three days. We left at dawn this morning to lay mines off the channel to Harwich harbour; a nest from which submarines, cruisers and destroyers buzz in and out like wasps. It will be ticklish work.

Torplexed

On the bottom.

Our mines are still with us, but so are our lives, which is something. We were approaching the appointed spot at 6 a.m. this morning, when without the slightest warning the track of a torpedo was seen streaking towards us about 50 yards on the starboard bow. Before Alten (who was on the bridge with me) could do more than press the diving alarm, the track met our ram. I breathed again, and was then reminded by an oath from Alten that the boat was diving.

It was evident that we had only been saved by the torpedo running deep under the cut-away part of our bow, otherwise!--well, the tangle of my affairs would have been easily straightened. Further procedure on the surface was suicidal, and we kept hydrophone patrol, twice hearing the motors of the enemy submarine. At the moment we are on the bottom waiting to come up and charge tonight, and lay our mines at dawn tomorrow.

On the bottom in 28 metres and feeling none too comfortable, as there would appear to be about a dozen destroyers overhead.

Last night, or rather early this morning, I participated in one of the most extraordinary incidents that I have ever heard of. It was pitch-black dark when I took over at 4 a.m., and a fresh breeze had raised a lumpy sea, which covered the bridge with spray. We were charging 400 amps on each, with the intention of laying one mine directly there was sufficient light to get a fix from some of the buoys which the English stick down all over the place here in the most convenient manner possible. If only one could believe they never shifted them. Alten says it never occurs to an Englishman to do a thing like that, but I'm not so sure. However, we were proceeding along at about five knots, crashing into the sea rather badly, when out of the black beastliness of the night I saw a shape close aboard on the port hand.

As I hesitated for a second as to my course of action, I was astounded to see a large submarine which must have been British, on an opposite course, not more than 25 metres away! This sounds absurd, but it really wasn't further. I'm not ashamed to confess that I was completely disorganized; it did not seem possible that the enemy was literally alongside me. I don't know how it struck the officer in the British boat, but I must give him credit for doing something first, for he fired a Very's white light straight at me as the two boats passed. It impinged on the hull, and in the flash I caught a photographic glimpse of his conning tower, on which was painted the letter E, followed by two numbers, of which one was a two I think, and the other a nine. By this time he was on my port quarter and rapidly disappearing; in a frenzy of rage I managed to get my revolver out, and whilst with the left hand I pressed the diving alarm, with the right hand I emptied the magazine in his direction. When we were down, Alten practically refused to believe me, which made me very pleased that in descending I had trod on a pair of hands which turned out to be his, as he had started up the ladder to the upper conning tower when he first heard the alarm.

I presume our opponent dived as well, but evidently he had put two and two together and used his aerial at some period, for when at dawn we poked a periscope up, a flotilla of destroyers appeared to be looking for something, which "something" was us, unless I am much mistaken; so we bottomed, where we have been ever since. The Hydroplane Operator keeps up a monotonous sing-song to the effect that "Fast running propellers are either receding or approaching." The crew are collected round the mine-tubes as I write, and are singing a lugubrious song, the refrain of which runs:

"Death for the Fatherland! Glorious fate,

This is the end that we gladly await."

Why will the seamen always become morbid when possible? And there is not a man amongst them who is not inwardly thinking of some beer-hall in Bruges, though I suppose that like their betters they have their romances of a tenderer kind.

The boat has been rolling about on the bottom in the most sickening manner the whole afternoon. We flooded P and Q to capacity, which gave her 50 tons negative, but it seems to have little effect in steadying her, and it is evident that a really heavy gale is running on top.

Surfaced at 10 p.m.; a very heavy sea running and impossible to do much more than heave to. This weather has one point in its favour and that is that the destroyers are driven in. It got steadily worse all night, and at midnight we lost our foremost wireless mast overboard; we have now (10 a.m.) been 48 hours without communication. At dawn we could see nothing to fix by; not a buoy in sight, nothing but an expanse of foam-topped short steep waves of dirty neutral-tinted water; how different to the great green and white surges of the broad Atlantic. Under these circumstances Alten decided to risk it and return without laying our mines; for once in a way I agreed with him, as it is better not to lay a minefield at all than dump one down in some unknown position which one may have to traverse oneself in the course of a month or so. We are now slowly, very slowly, struggling back to Zeebrugge.

A green sea came down the conning tower today, and everything in the boat is damp and smelly and beastly. The propellers race at frequent intervals and the whole boat shudders--I feel miserable. Alten has started to drink spirits; he began as soon as we decided to go back. He will be incapable by tonight, and it means that I shall have to take her in.

What hell this is, sitting in sodden clothes, with the stench of four days' living assaulting the nostrils, and a motion of the devil; the glass is very low and is slowly rising, so that I suppose it will blow harder soon, though it is about force eight at present.

I wonder what Zoe will have written in reply to my note. When I think of what I rejected and compare it with my beast-like existence here, I can hardly believe that I behaved as I did--what would I not give now to be transported back to the forest! At this rate of progress we shall take another 24 hours. I wonder if I can knock another half-knot out of her without smashing her up.

The extraordinarily violent motion has upset the Anschutz.[14] The bearing cone of the stabilizing gyro has cracked, and the master compass began to wander off in circles. I was just resting for an hour or two, wedged up on a wet settee with coats equally wet, when her heavy pitching changed to a wallowing roll, and I heard the pilot, who was on watch, cursing down the voice-pipe, as we had sagged off our course.

I heard the voice of the helmsman querulously maintain that he was steering his course by Anschutz, so I got up and gingerly clawed my way into the control room, where I found by comparing Anschutz with magnetic that the former had gone to hell, the reason being obvious, as the stabilizer was exerting a strongly biased torque. I stopped the Anschutz and asked the pilot to give the helmsman a steady by magnetic.

As we staggered back to our course I heard a thud in the wardroom, and on returning to my settee found that Alten had rolled out of his bunk, where he was lying in a drunken stupor, and that he was face downwards, sprawling on the deck, half his face in the broken half of a dirty dish which had fallen off the table whilst I was having tea. As I couldn't let the crew see him like this, I was obliged to struggle and get him back into his bunk. He was like a log and absolutely incapable of rendering me any assistance, though he did open his eyes and mutter once or twice as I lifted him up, trunk first and then his legs. He stank of spirits and I hated touching him. Lord! what a truly hoggish man he is; yet I cannot help envying him his oblivion to these surroundings.

Arrived in, this afternoon.

Alten quite slept off his drink, and was offensively sarcastic as I worked on the forepart with wires, getting her into the shelters alongside the mole.

---

[14] 14. Gyroscopic compass.--ETIENNE.

I hastened up to Bruges, and in the Mess heard several items of news and found two letters. The first, in a well-known handwriting, I opened eagerly, but received a chill of disappointment when I read its single line.

"I am here when you want me.--Z."

So she thinks to break my resolution! No! I am stronger than she, and, now that I know she loves me, I can and will bend her to my will. Even now, at this distance of time, I can hardly understand my conduct the other day. I must have been given the strength of ten. I feel that I could not do it again; had she hesitated a second longer at the door--well, I can hardly say what I would have done.

It is my duty to do so, for her sake and my own. But I know my weakness, and in this fact lies my strength. Cost what it may, I shall not permit myself to go near her until she yields.

The second letter gave me a great surprise. It was from Rosa. She has passed some examination, and is coming here of all places as a Red Cross nurse. She says she is looking forward to going round a U-boat! She assumes a good deal, I must say, still, I suppose I must be polite to her; but why the deuce does she sign herself "Yours, Rosa?" She's not mine, and I don't want her; it seems funny to me that I once thought of her vaguely in that sort of way. Now, I feel rather disturbed that she is coming here, though I don't quite see why I should worry, and yet I wonder if it is a coincidence her coming to Bruges?

I'm almost inclined to think it isn't. After all, every girl wants to get married, and without conceit my family, circumstances and, in the privacy of the pages of this journal I may add, my personal appearances, are such as would appeal to most girls--except Zoe, apparently!

I'll have to be on my guard against Miss Rosa.

I heard today that I am likely to be appointed to the periscope school in a few weeks' time, and meanwhile I am to be attached as supernumerary to the operations division on old Max's staff.

The work here is most interesting. I feel glad that I am one of the spiders weaving the web for Britain's destruction.

The impasse with Zoe still continues, and my peace of mind has been still further disturbed by the actual arrival of Rosa. She rang me up within twelve hours of her arrival, and, of course, I was obliged to call. That was the day before yesterday. Rosa is at the No. 3 Hospital here, and was horribly effusive. Some people would, I suppose, call her good-looking, but to me, with my mind's-eye in perpetual contemplation of my darling Zoe, Rosa looked like a turnip. Her first movement after the preliminary greetings was to offer me a cigarette! I then noticed that her fingers were stained with nicotine, unpleasant in a man, disgusting in a woman.

Her nose was shiny and greasy--horrible. After a little talk she volunteered the statement that yesterday was her afternoon off, and she was simply longing to have tea in the gardens. I endeavoured to make some feeble excuse on the grounds of the weather being unsuitable, but I am no good at these social lies, and I was eventually obliged to promise to take her there. I was the more annoyed in that her main object was obviously to be seen walking with a U-boat officer.

Accordingly, yesterday, I found myself walking about with her at my side. My feelings can better be imagined than described when I suddenly saw Zoe, accompanied by Babette, in the distance. I hastily altered course, and pray she didn't see me.

In the course of the afternoon Rosa had the impertinence to say that at Frankfurt they were saying that I was interested in a beautiful widow at Bruges, and could she (Rosa) write and say I was heart-whole, or else what the girl was like. I'm afraid that I lost my temper a little, and I told Rosa she could write to all the busybodies at home and tell them from me to go to the devil. These women in the home circle, and especially aunts, are always the same; firstly, they badger one to get married, and then if they think one is contemplating such a step they are all agog to find out whether she is suitable!

Three more boats, two of which are U.C.'s, are overdue. It is distinctly unpleasant not knowing how or where they go, though the U.B. boat (Friederich Althofen) made her incoming position the day before yesterday as off Dungeness, so it looks as if the barrage at Dover which got Weissman has got Althofen as well. I wonder what new devilry they have put down there.

How one wishes that in 1914, instead of seeking the capture of Paris, we had realized the importance of the Channel Ports to England, and struck for them! It would not have been necessary to strike even in September, 1914. We could have walked into them. Dunkirk, at all events, should have been ours; however, we must do the best with things as they are, not that I would consider it too late even now to make a big push for the French coast.

It would seem, as a matter of fact, that all the pushing is to be at the other end of the line, in the Verdun sector, from the rumours I hear, though I should have thought once bitten twice shy in that quarter.

Saw Zoe again in the distance, and I think she saw me; at all events she turned round and walked away.

This girl whom I cannot, and would not if I could, obliterate from my thoughts, is causing me much worry. She shows no sign of giving in, and I for one intend to be adamant. I shall defeat her in time. The male intellect is always ultimately victorious, other things being equal. I was reading Schopenhauer on the subject last night. What a brain that man had, though I confess his analysis of the female mentality is so terribly and truthfully cruel that it jars on certain of my feelings.

Zoe's resolution in this conflict, this sex war one might call it, only adds to her charm in my eyes; she is, I feel, a worthy mate for me, both intellectually and physically, and she shall be mine--I have decided it.

Met Rosa today at old Max's house, where I went to pay a duty call. Her Excellency is as forbidding a specimen of her sex as any I have ever met. She quite frightened me, and in the home circle the old man seemed quite subdued.

I escorted Rosa home, and on the way to her hospital she gave me a great surprise, as after much evasive talk she suddenly came out with the news that she was engaged to Heinrich Baumer, of U.C.23. I was quite taken aback, and will frankly confess that not so very long ago I imagined, evidently erroneously, that she was disposed to let her affections become engaged in another quarter. However, I was really very glad to hear this news, and congratulated her with genuine feeling.

The knowledge that she was a promised woman quite altered my feelings towards her, and before I quite meant to, I had told her a considerable amount about Zoe. It gave me much relief to be able to unburden myself, and confide my difficulties elsewhere than in the pages of this journal. I have asked the girl to tea tomorrow.

A vile air raid last night. British machines, of course. They seemed determined to get over the town, and from 1 a.m. to 3 a.m. relays of machines (of which not one was shot down) attacked us. The din was tremendous, and all sleep was out of the question.

Morning revealed surprisingly little damage, as is often the case in these big raids, whereas a few bombs from a chance machine often work havoc. I was down at 50 B.C. aerodrome this morning, and heard that as soon as the moon suits we are going to make Dunkirk sit up as retaliation for last night's efforts. There were also rumours of big attacks impending on London as soon as the new type of Gothas are delivered. That will shake the smug security of those cursed islanders.

Rosa came to tea, and afterwards I told her more about Zoe, and as I expect any day to be appointed to the periscope school at Kiel, I asked Rosa to try and effect an introduction to Zoe, and do what she could for me. Rosa gave me the impression that she was somewhat surprised that I should have had any difficulty with Zoe (of course I had not told her of the shooting-box scene). Rosa evidently thinks any woman ought to be honoured....

Perhaps I was not so far wrong in my surmises as to Rosa's previous inclinations--I wonder; at any rate she will undoubtedly make Baumer a good wife, and she will probably be very fruitful and grow still fatter and housewifely. She is of a type of woman appointed by God in his foresight as breeders. Zoe, my adorable one, will probably not take kindly to babies.

I am ordered to report myself at Kiel by next Monday. I am terribly tempted to ring up Zoe on the telephone before I leave: it seems dreadful to leave her without a word; but at the same time I feel that she would interpret this as a sign of weakness on my part--as indeed it would be. I must be firm, for strength of mind pays with women, even more than with men.

At Kiel.

I left Bruges without a word either to or from my obstinate darling.

It is torture being away from her. I had thought that when I was here and not exposed to the temptation of going round and seeing her, that it would be easier; it is not. I long to write, and how I wonder whether she is feeling it as I do. I have read somewhere that a woman's passion once aroused is more ungovernable than a man's. That her whole being cries aloud for me cannot be doubted, and if the above statement is true what inflexibility of will she must be showing--it almost makes me fear--but no, I will defeat her in this strange contest, and she shall be my wife.

The work here is strenuous, and the grass does not grow under one's feet. The course for commanding officers lasts four weeks, and terminates in an exceedingly practical but rather fearsome test--i.e., they have six steamers here camouflaged after the English fashion with dazzle painting, and these six steamers, protected by launches and harbour defence craft, steam across Kiel Bay in the manner of a convoy. The officer being examined has to attack this group of ships in one of the instructional submarines, and in three attacks he must score at least two hits, or else, in theory, he is returned to general service in the Fleet.

Fortunately at the moment I hear that owing to recent losses they are distinctly on the short side where submarine officers are concerned, so they'll probably make it easy when I do my test.

⊕ ⊕ ⊕

I see I have written nothing here for a fortnight; this is due to two causes: Firstly, I have been so extraordinarily busy, and, secondly, I have been most depressed through a letter I received from Fritz. It contained two items of bad news.

In the first place, I heard for the first time of the tragedy of Heinrich Baumer's boat, and to my astonishment Fritz tells me that Rosa and another girl were in her when she was lost!

It appears that she was to go out for a couple of hours' diving off the port as a matter of routine after her two months' overhaul. She went out at 10 a.m., and was sighted from the signal station at the end of the mole at 11.30, when almost immediately afterwards there was an explosion and she disappeared. Motor-boats were quickly on the scene, but only debris came to the surface. Divers were sent down, and reported that she was in ten metres of water completely shattered. It is assumed, for lack of other explanation, that she struck a chance drifting mine which was moving down the coast on the tide.

Meanwhile Rosa and another sister were missing from the hospital, and after forty-eight hours someone put two and two together and started investigations. It has been ascertained that Baumer motored down from Bruges after breakfast, and that in the car were two figures taken to be sailors, as they were muffled up in oilskins. This fact was noted by the control sentries, as, though the day was showery, it was not raining hard. Other scraps of evidence unite in showing that these were the two girls who had apparently induced Baumer to take them out for a dive as a treat.

What a tragedy! However, it must have been quite instantaneous. Poor Rosa, with all her vanities about war work, to think that the war would claim her like that![15]

Fritz added that old Max is almost off his head with rage over the whole business, and it is difficult to say whether he is more angry over Baumer and the boat being lost, or over the fact that Baumer being dead he is unable to administer those "disciplinary actions" in which he delights.

Great excitement here, as the day after tomorrow His Imperial Majesty the Kaiser and Hindenburg are due to pay Kiel a surprise visit. We are to be inspected and addressed. Tremendous preparations are going on.

His Majesty, accompanied by the great Field-Marshal, inspected us this morning, and made a fine speech, of which we have been given printed copies. I shall frame mine and hang it in my boat, if I get a command.

I transcribe it:

"Officers and men of the U-boat service:

"In the midst of the anxious moments in which we live I have determined to make time to come and witness in my own person the labours of those on whom I and the Fatherland rely. Fresh from the great battles on the West which are gnawing at the vitals of our hereditary enemies, I come to those whose glorious mission

---

[15] It is known that a boat with women on board was lost whilst exercising off Zeebrugge in the Spring of 1917. This would appear to be the boat in question.--ETIENNE.

it will be to strike relentlessly at our most deadly and cunning enemy--cursed Britain. God is on our side and will protect you at sea for, in the striking at the nation which openly boasts that it aims at starving our women and children, you are engaged on a mission of undoubted holiness.

"You must sink and destroy even as of old the Israelites smote and destroyed the alien races.

"To the officers I would particularly say, my person is your honour, and I am your supreme chief. From my hands you will receive honour, and from my hands will proceed just punishment for the unhappy ones who fail in their duty.

"To the men I would say, trust and obey your officers as you would your God. Officers and men! In you, your Kaiser and Fatherland place their trust--let neither be disappointed!"

After his address, His Majesty graciously spoke a few words to individuals, of whom I had the signal honour of being one. I felt that I was in the presence of an Emperor. His gestures, his eyes, his voice, impressed me as belonging to a man born to command and to fill high places. The Field-Marshal never opened his mouth. I understand from his A.D.C. that he rarely speaks in public.

The Colonel is KILLED! When I think about it, I am so excited I can hardly write!

I heard the great news last night, quite by accident. I was sitting in the Mess after dinner, and picked up Die Woche, and glancing at the pictures, I suddenly saw the portrait of Colonel Stein, of the Brandenburgers, killed on the 7th instant near Ypres. I recognized the ugly and bloated face immediately from the photograph of him which she had once shown me.

My first impulse was to send her a wire, but, on thinking matters over, I decided that it would be difficult to put all my thoughts into the curt sentences of a telegram, and, further, that as all wires are doubtless examined at the Main Post Office at Bruges, it might lead to trouble, so I wrote her a letter. This, in a way, has been an exhibition of weakness on my part, as I had promised myself that I would not take the first step in reopening communication; but I feel that the fortunate death of Stein has completely altered the case. I told her in the letter that I realized that I had made mistakes, but that if she still loved me with half the strength that I loved her, then a telegram to me would make me the happiest of men.

I wrote that yesterday, but have had no wire. Perhaps, like me, she distrusts telegrams and prefers letters.

A long letter from Zoe: an accursed letter--an abominable letter--a damnable letter; she still refuses to marry me. I leave for Bruges tonight on forty-eight hours' special leave.

⊕   ⊕   ⊕

Kiel, 17th

I hate Zoe, she has broken my heart.

After her preposterous letter of the 14th, I decided that in a matter which so closely affected my happiness no stone ought to remain unturned to ensure a satisfactory solution of the problem, so I determined to have a personal interview. I arrived at Bruges after tea and went at once to the flat. I tackled her immediately on the subject of her letter, and told her that naturally I understood that a decent interval must elapse before we married; but, granted this fact, I told her that I failed to see what prevented our marriage.

A most unpleasant and harrowing scene ensued, the details of which form such painful recollections that I really cannot write them down here, though in the passage of months I have acquired the habit of writing in the pages of this journal with the same freedom as I would talk to that wife whom I had hoped to possess. She maintained an obstinate silence when I urged her to give me at least some tangible reason as to why she would not marry me. She contented herself and maddened me by reflecting in a kind of monotone: "I love you, Karl! and am yours, but I cannot marry you."

I could have beaten her till she was senseless, but I had enough sense to realize that with Zoe, whose resolution, considering she is a woman, amazes me, force is not the best method. As I continued to press her (time was important: had I not journeyed far to see her?), those glorious eyes of hers, which I love and whose power I dread, filled with tears. I was a brute! I was heartless! I was inconsiderate! I could not love her! I was cruel! And I know not what other accusation crushed me down. Broken-hearted and dispirited, I told her to choose there and then.

She collapsed on to a sofa in a storm of tears, and after a severe mental struggle I took the only possible course, and leaving the room--left her for ever. I have resumed my service life determined to cast her out from my mind.

I will not deceive myself: it will be hard. Love and Logic are deadly enemies, but Logic must and shall prevail. Though I have seen her for the last time, I cannot escape the net of fascination which the girl has thrown over me. Perhaps in the course of time I shall slowly emerge and free myself from its entanglements. At present I hate her for this blow she has dealt me, and yet, O Zoe! my darling, how I long to be with you!

Today I went through my final test for qualification as U-boat commander.

At 9 a.m. I proceeded to sea in command of the U.11, one of the instructional boats here. We proceeded out into Kiel Bay. On board and watching my every movement was a committee consisting of a commander and two lieutenant-commanders.

On arrival at the entrance lightship, I was ordered to attack a convoy of camouflaged ships which were just visible about fifteen kilo-metres away off the Spit Bank. I had a very shrewd idea as to the course they would steer, and on coming up for my final observation I found myself in an excellent position, 1,000 metres on the bow of the leading ship. The rest was easy. I gave the leader the two bow torpedoes, and, turning sixteen points, fired my stern tube at the third ship of the line. Two hits were obtained, and I returned to harbour well pleased with myself. There is not the slightest chance of having failed to qualify.

My confidence in myself was not misplaced; I heard today that I am on the command list, and anticipate in a few days being appointed to a boat. I wonder which craft I shall get?

I met the A.D.C. to the Chief of the Staff at the school, at the gardens, and in conversation with him discovered that he had heard that three boats were being detached from the Flanders flotilla for an unknown destination. This has given me an idea, for I feel that I can never return to Bruges, and I was rather dreading being appointed to one of the boats there. I have dropped a line to Fritz Regels, who is on old Max's staff, and told him that

I do not wish to return to Bruges, and I further hinted that I understood a detached squadron was proceeding somewhere, and, as far as I was concerned, the further the better, if I could get into it.

I have tried the night life at this place at the Mascotte and Trocadero,[16] in order to forget, but it is a poor consolation.

⊕   ⊕   ⊕

A letter from Fritz, saying that he has an idea that Korting's boat would suit me, though he could not of course give me further details in a letter; however, he informs me positively that I shall not be at Bruges.

On the strength of this I have wired to Fritz, and asked him to try and fix up an exchange between me and Korting, provided the latter is agreeable and the people in Max's office have no objection. I have a recollection that Korting's boat is one of the U.40--U.60 class, which would suit me admirably, and, as for destination, I care not where it is, provided only that it be far from Bruges.

At sea.

I have quite neglected my poor old journal for several weeks. But I have passed through an extraordinarily busy period.

It was approved that I should relieve Korting, whose boat, the U.59, I discovered to be refitting at Wilhelmshaven. I was very pleased not to go back to Bruges, though as we steam steadily north at this moment I cannot escape a sense of deep disappointment that upon my return from this trip I shall not enjoy as of old the fascination of Zoe. But I shall have plenty of time to get accustomed to this idea, for this is no ordinary trip.

We are bound for the North Cape and Murman Coast, where we remain until well into the cold weather--at any rate, for three months. Our mission is to work off that fogbound and desolate coast, and attack the constant stream of traffic between England and Archangel. There are two other boats besides ourselves on the job, but we shall all be working far apart.

---

[16] Two well-known cabarets at Kiel.--ETIENNE.

Our first billet is off the North Cape. In order to save time, we are to be provisioned once a month in one of the fjords. I don't imagine the Admiralty will have any difficulty in getting supplies up to us, as at the moment we are off the Lofotens, and we actually have not had to dive since we left the Bight!

There seems to be nothing on the sea except ourselves. Where is the much vaunted and impenetrable web of blockade which the English are supposed to have spread around us? And yet many raw materials are getting very short with us. I see that in this boat they have replaced several copper pipes with steel ones during her refit, and this will lead to trouble unless we are careful--steel pipes corrode so badly that I never feel ready to trust them for pressure work.

The truth about the blockade is that it is largely a paper blockade, yet not ineffective for all that. Unfortunately for us, the damned English and their hangers-on control the cables of the world, and hence all the markets, and I don't suppose, to take the case of copper, that a single pound of it is mined from the Rio Tinto without the British Board of Trade knowing all about it. The neutral firms simply dare not risk getting put on to the British Black List; it means ruination for them. And then all these dollar-grabbing Yankees, enjoying all the advantages of war without any of its dangers--they make me sick.

This seems a most profitable job. I have only been up seven days, but I've bagged four steamers, all by gun-fire, and all fat ships, brimful of stuff for the Russians. My practice has been to make the North Cape every day or two to fix position, as the currents are the most abnormal in these parts, and I should say that the "Sailing Directions Pilotage Handbook" and "Tidal Charts" were compiled by a gentleman at a desk who had never visited these latitudes.

At the moment I am standing well out to sea, as the immediate vicinity of the North Cape has become rather unhealthy.

Yesterday afternoon (I had sunk number four in the morning, and the crew were still pulling for the coast) four British trawlers turned up. These damned little craft seem to turn up wherever one goes. I longed to have a bang at them with my gun, but, apart from the uncertainty as to what they carried in the way of armament, I have strict orders to avoid all that sort of thing, so I dived and steamed slowly west, came up at dusk and proceeded to charge up my batteries.

These U.6O's are excellent boats, and I am very lucky to get one so soon. I suppose Korting, being a married man, wants to stay near his wife. I cannot write that word without painful memories of Zoe and idle thoughts of what might have been. Well, perhaps it is for the best. I am not sure that a member of the U-boat service has the right to get married in war-time, for unless he is of exceptional mentality it must affect his outlook under certain circumstances, though I think I should have been an exception here. Then the anxiety to the woman must be enormous; as every trip comes round a voice must cry within her, this may be the last. The contrast between the times in harbour and the trips is so violent, so shattering and clear cut.

With a soldier's wife, she merely knows that he is at the front; with us, at 8 p.m. one may be kissing one's wife in Bruges, and at 6 a.m. creeping with nerves on edge through the unknown dangers of the Dover Barrage-- but I have strayed from what I meant to write about--my first command and her crew.

The quarters in this class are immensely superior to the U.C.-boats. Here I have a little cabin to myself, with a knee-hole table in it. My First Lieutenant, the Navigator and the Engineer have bunks in a room together, and then we have a small officers' mess. On this job up here, as we are not to return to Germany for supplies, and, consequently, I should say we may have to live on what we can get out of steamers, I don't propose to use my torpedoes unless I meet a warship or an exceptionally large steamer.

The gun's the thing, as Arnauld de la Perrière has proved in the Mediterranean; but half the fellows won't follow his example, simply because they don't realize that it's no use employing the gun unless it is used accurately, and good shooting only comes after long drill. I have impressed this fact on my gun crew, and particularly the two gun-layers, and I make Voigtman (my young First Lieutenant) take the crew through their loading drill twice a day, together with practice of rapid manning of the gun after a "surface" or rapid abandonment of the gun should the diving alarms sound in the middle of practice. I have also impressed on Voigtman that I consider that he is the gun control officer, and that I expect him to make the efficient working of the gun his main consideration.

As regards the crew, they are the usual mixed crowd that one gets nowadays: half of them are old sailors, the others recruits and new arrivals from the Fleet. My main business at the moment is to get the youngsters into shape, and for this purpose I have been doing a number of crash dives. It also gives me an opportunity of getting used to the boat's peculiarities under water. She seems to have a tendency to become tail-heavy, but this may be due to bad trimming.

Voigtman has been in U.B.43 for nine months, and seems a capable officer. Socially, I don't think he can boast of much descent, but he has no airs, and treats me with pleasing respect, apart from service considerations.

A very awkward accident took place this morning, which resulted in severe injury to Johann Wiener, my second coxswain.

A party of men under his direction were engaged in shifting the stern torpedo from its tube, in order to replace it with a spare torpedo, as I never allow any of my torpedoes to stay in the tube for more than a week at a time owing to corrosion. The torpedo which had been in the tube had been launched back and was on the floor plates. The spare torpedo, destined for the vacant tube, was hanging overhead, when without any warning the hook on the lifting band fractured, and the 1,000 kilogrammes' mass of metal crashed down.

Wonderful to relate, no one was killed, but two men were badly bruised, and Wiener has been very seriously injured. He was standing astride the spare torpedo, and his right leg was extremely badly crushed, mostly below the knee. Unfortunately it took about ten minutes to release him from his position of terrible agony. I should have expected him to faint, but he did not. His face went dead white, and he began to sweat freely, but otherwise endured his ordeal with praiseworthy fortitude.

I am now confronted with a perplexing situation. I cannot take him back to Germany; I cannot even leave my station and proceed south to any of the Norwegian ports. If I could find a neutral steamer with a doctor on board, I would tranship him to her; but the chances of this God-send materializing are a thousand to one in these latitudes. If I sighted a hospital ship I would close her, but as far as I know at present there are no hospital ships running up here. The chances of outside assistance may therefore be reckoned as nil. Wiener's hope of life depends on me, and I cannot make up my mind to take the step which sooner or later must be taken--that is to say, amputation.

It is a curious fact, but true, nevertheless, that although, as a result of the war, men's lives, considered in quantity, seem of little importance, when it comes to the individual case, a personal contact, a man's life assumes all its pre-war importance. I feel acutely my responsibility in this matter. I see from his papers that he is a married man with a family; this seems to make it worse. I feel that a whole chain of people depend on me.

Since I wrote the above words this morning, Wiener has taken a decided turn for the worse.

I have been reading the "Medical Handbook," with reference to the remarks on amputation, gangrene, etc., and I have also been examining his leg. The poor devil is in great pain, and there is no doubt that mortification has set in, as was indeed inevitable. I have decided that he must have his last chance, and that at 8 p.m. tonight I will endeavour to amputate.

Midnight.

I have done it--only partially successful.

Last night, in accordance with my decision, I operated on Wiener. Voigtman assisted me. It was a terrible business, but I think it desirable to record the details whilst they are fresh in my memory, as a Court of Inquiry may be held later on. Voigtman and I spent the whole afternoon in the study of such meagre details on the subject as are available in the "Medical Handbook." We selected our knives and a saw and sterilized them; we also disinfected our hands.

At 7.45 I dived the boat to sixty metres, at which depth the boat was steady. We had done our best with the wardroom-table, and upon this the patient was placed. I decided to amputate about four inches above the knee, where the flesh still seemed sound. I considered it impracticable to administer an anaesthetic, owing to my absolute inexperience in this matter.

Three men held the patient down, as with a firm incision I began the work. The sawing through the bone was an agonizing procedure, and I needed all my resolution to complete the task. Up to this stage all had gone as well as could be expected, when I suddenly went through the last piece of bone and cut deep into the flesh on the other side. An instantaneous gush of blood took place, and I realized that I had unexpectedly severed the popliteal artery, before Voigtman, who was tying the veins, was ready to deal with it.

I endeavoured to staunch the deadly flow by nipping the vein between my thumb and forefinger, whilst Voigtman hastily tried to tie it. Thinking it was tied, I released it, and alas! the flow at once started again; once more I seized the vein, and once again Voigtman tried to tie it. Useless--we could not stop the blood. He would undoubtedly have bled to death before our eyes, had not Voigtman cauterized the place with an electric soldering-iron which was handy.

Much shaken, I completed the amputation, and we dressed the stump as well as we could.

At the moment of writing he is still alive, but as white as snow; he must have lost litres of blood through that artery.

9 p.m.

Wiener died two hours ago. I should say the immediate cause of death was shock and loss of blood. I did my best.

⊕   ⊕   ⊕

We have been out on this extended patrol area seven days, but not a wisp of smoke greets our eyes.

Nothing but sea, sea, sea. Oh, how monotonous it is! I cannot make out where the shipping has got to. Tomorrow I am going to close the North Cape again. I think everything must be going inside me. I am too far out here.

The North Cape bears due east. Nothing afloat in sight. Where the devil can all the shipping be? In ten days' time I am due to meet my supply ship; meanwhile I think I'll have to take another cast out, of three hundred miles or so.

Nothing in sight, nothing, nothing.

The barometer falling fast and we are in for a gale. I have decided to make the coast again, as I don't want to fail to turn up punctually at the rendezvous.

In the Standarak-Landholm Fjord--thank heavens.

Heavens! we have had a time. We were still two hundred and fifty miles from the coast when we were caught by the gale. And a gale up here is a gale, and no second thoughts about it. To say it blew with the force of ten thousand devils is to understate the case. The sea came on to us in huge foaming rollers like waves of attacking infantry intent on overwhelming us.

We struggled east at about three knots. But she stuck it magnificently. Low scudding clouds obscured the sky and came like a procession of ghosts from the north-east. Sun observations were impossible for two reasons. Firstly, no one could get on deck; secondly, there was no visible sun. This lasted for three days, at the end of which time we had only the vaguest idea as to where we were.

The gale then blew out, but, contrary to all expectations, was succeeded by a most abominable fog, thick and white like cotton-wool. These were hardly ideal conditions under which to close a rocky and unknown coast, but it had to be done. The trouble was that it was entirely useless taking soundings, as the twenty-metre depth-line on the chart went right up to the land. We crept slowly eastwards, till, when by dead reckoning we were ten miles inside the coast, the Navigator accidentally leant on the whistle lever; this action on his part probably saved the ship, as an immediate echo answered the blast. In an instant we were going full-speed astern. We altered course sixteen points and proceeded ten miles westerly, where we lay on and off the coast all night, cursing the fog.

Next day it lifted, and we spent the whole time trying to find the entrance to the S. Landholm Fjord. The coast appeared to bear no resemblance to the chart whatsoever. The cliffs stand up to a height of several hundred metres, with occasional clefts where a stream runs down. There are no trees, houses, animals, or any signs of life, except sea birds, of which there are myriads. The Engineer declares he saw a reindeer, but five other people on deck failed to see any signs of the beast.

After hours of nosing about, during which my heart was in my mouth, as I quite expected to fetch up on a pinnacle rock, items which are officially described in the Handbook as being "very numerous," we rounded a bluff and got into a place which seems to answer the description of S. Landholm. At any rate, it is a snug anchorage, and here I intend to remain for a few days, and hope for my store-ship to turn up. I've posted a daylight look-out on top of the bluff; it would be very awkward to be caught unawares in this place, which is only about 150 metres wide in places. I'm taking advantage of the rest to give the crew some exercises and execute various minor repairs to the Diesels.

Yesterday we fought what must be one of the most remarkable single-ship actions of the war.

At 9 a.m. the look-out on the cliffs reported smoke to the northward. I got the anchor up and made ready to push off, but still kept the look-out ashore. At 9.30 he reported a destroyer in sight, which seemed serious if she chose to look into my particular nook.

At any rate, I thought, I wouldn't be caught like a rat, so I got my look-out on board--a matter of ten minutes--and then proceeded out, trimmed down and ready for diving. When I drew clear of the entrance I saw the enemy distant about a thousand metres. I at once recognized her as being one of the oldest type of Russian torpedo boats afloat. When I established this fact, a devil entered into my mind, and did a most foolhardy act.

I decided that I would not retreat beneath the sea, but that I would fight her as one service ship to another. When I make up my mind, I do so in no uncertain manner--indecision is abhorrent to me--and I sharply ordered, "Gun's Crew--Action."

I can still see the comical look of wonderment which passed over my First Lieutenant's face, but he knows me, and did not hesitate an instant. We drilled like a battleship, and in sixty-five seconds--I timed it as a matter of interest--from my order we fired the first shot. It fell short.

Extraordinary to relate, the torpedo boat, without firing a gun, put her helm hard over, and started to steam away at her full speed, which I suppose was about seventeen knots. I actually began to chase her--a submarine chasing a torpedo boat! It was ludicrous.

With broad smiles on their faces, my good gun's crew rapidly fired the gun, and we had the satisfaction of striking her once, near her after funnel, but it did no vital damage, as a few minutes afterwards she drew out of range! What a pack of incompetent cowards! They never fired a shot at us. I suppose half of them were drunk or else in a state of semi-mutiny, for one hears strange tales of affairs in Russia these days.

The whole incident was quite humorous, but I realized that I had hardly been wise, as without doubt the English will hear of this, and these trawlers of theirs will turn up, and I'm certainly not going to try any heroics with John Bull, who is as tough a fighter as we are.

Meanwhile, what of the supply ship, for I'm supposed to meet her here, and it's already twenty-four hours since yesterday's epoch-making battle and I expect the English any moment.

My doubts were removed for me since I received special orders at noon by high-power wireless from Nordreich, and on decoding them found that, for some reason or other, we are ordered to proceed to Muckle Flugga Cape, and thence down the coast of Shetlands to the Fair Island Channel, where we are directed to cruise till further orders. Special warning is included as to encountering friendly submarines. It appears to me that a special concentration of U-boats is being ordered round about the Orkneys, and that some big scheme is on hand.

We are now steering south-westerly to make Muckle Flugga, which I hope to do in four days' time if the weather holds. These Northern waters have proved very barren of shipping in the last few weeks, and this fact, coupled with the approaching winter weather, which must be fiendish in these latitudes, makes me quite ready to exchange the Archangel billet for the work round the Orkneys and Shetlands, though this is damnable enough in the winter, in all conscience.

There is only one fly in the ointment, and that is that this premature return to North Sea waters might conceivably mean a visit to Zeebrugge, though this class are not likely to be sent there. Though it is many weeks since I left Zoe, I have not been able to forget her. I continually wonder what she is doing, and often when I am not on my guard she wanders into my thoughts. Whilst I am up here, it does not matter much, except that it causes me unhappiness, but if I found myself at Bruges it would be very hard. However, I don't suppose I shall ever see her again.

Sighted Muckle Flugga this morning, and shaped course for Fair Island.

⊕ ⊕ ⊕

Oh! what a hell I have passed through. I can hardly realize that I am alive, but I am, though whether I shall be tomorrow morning is doubtful--it all depends on the weather, and who would willingly stake their life on North Sea weather at this time of the year? Curses on the man who sent us to the Fair Island Channel. Where the devil is our Intelligence Service? If we make Flanders I have a story to tell that will open their eyes, blind bats

that they are, luxuriating in the comfort of their fat staff jobs ashore.

The Fair Island Channel is an English death-trap; it stinks with death. By cursed luck we arrived there just as the English were trying one of their new devices, and it is the devil. Exactly what the system is, I don't quite know, and I hope never again to have to investigate it.

For forty-seven, hours we have been hunted like a rat, and now, with the pressure hull leaking in three places, and the boat half full of chlorine, we are struggling back on the surface, practically incapable of diving at least for more than ten minutes at a time. Even on the surface, with all the fans working, one must wear a gas mask to penetrate the fore compartment. Oh! these English, what devils they are!

Here is what happened:

Fair Island was away on our port beam when we sighted a large English trawler, which I suspected of being a patrol. To be on the safe side, I dived and proceeded at twenty metres for about an hour.

At 5 p.m. (approximately) I came up to periscope depth to have a look round, but quickly dived again as I discovered a trawler, steering on the same course as myself, about a thousand metres astern of me. This was the more disconcerting, as in the short time at my disposal it seemed to me that she was remarkably similar to the craft I had seen in the afternoon, and yet this hardly seemed likely, as I did not think she could have sighted me then.

On diving, I altered course ninety degrees, and proceeded for half an hour at full speed, then altered another ninety degrees, in the same direction as the previous alteration, and diving to thirty metres I proceeded at dead slow. By midnight I had been diving so much that I decided to get a charge on the batteries before dawn; I also wanted to be up at 1 a.m. to make my position report.

I surfaced after a good look round through the right periscope, which, as usual, revealed nothing. I had hardly got on the bridge, when a flash of flame stabbed the night on the starboard beam and a shell moaned just overhead.

I crash-dived at once, but could not get under before the enemy fired a second shot at us, which fortunately missed us. As we dived I ordered the helm hard a starboard, to counteract the expected depth-charge attack. We must have been a hundred and fifty metres from the first charge and a little below it, five others followed in rapid succession, but were further away, and we suffered no damage beyond a couple of broken lights. The situation was now extremely unpleasant. I did not dare venture to the surface, and thus missed my 1 a.m. signal from Headquarters. I wanted a charge badly, and so proceeded at the lowest possible speed. At regular intervals our enemy dropped one depth-charge somewhere astern of us, but these reports always seemed the same distance away.

At dawn I very cautiously came up to periscope depth, and had a look. To my consternation I discovered our relentless pursuer about 1,500 metres away on the port quarter. In some extraordinary manner he had tracked us during the night. I dived and altered course through ninety degrees to south.

At 9 a.m. a tremendous explosion shook the boat from stem to stern, smashing several lights, and giving her a big inclination up by the bow. As I was only at twenty metres I feared the boat would break surface, and our enemy was evidently very nearly right over us. I at once ordered hard to dive, and went down to the great depth of ninety-five metres.

A series of shattering explosions somewhere above us showed that we were marked down, and we were only saved from destruction by our great depth, the English charges being set apparently to about thirty metres.

At noon the situation was critical in the extreme. My battery density was down to 1,150, the few lamps that I had burning were glowing with a faint, dull red appearance, which eloquently told of the falling voltage and the dying struggles of the battery.

The motors with all fields out were just going round. The faces of the crew, pallid with exhaustion, seemed of an ivory whiteness in the dusky gloom of the boat, which never resembled a gigantic and fantastically ornamental coffin so closely as she did at that time.

The air was fetid. I struck a match; it went out in my fingers. The slightest effort was an agony. I bent down to take off my sea-boots, and cold sweat dropped off my forehead, and my pulse rose with a kind of jerk to a rapid beating, like a hammer. I left one sea-boot on.

At 1 p.m. a deputation of the crew came aft, and in whispered voices implored me to surface the boat and make a last effort on the surface. A muffled report, as our implacable enemy dropped a depth-charge somewhere astern of us, added point to the conversation, and showed me that our appearance on the surface could have but one end.

At 3 p.m. the second coxswain, who was working the hydroplanes, fell off his stool in a dead faint.

At 3.30 p.m. the supreme crisis was reached: two more men fainted, and I realized that if I did not surface at once I might find the crew incapable of starting the Diesels.

At the order "Surface," a feeble cheer came from the men.

We surfaced, and I dragged myself-up to the conning tower. Luckily we started the Diesels with ease, and in a few minutes gusts of beautiful air were circulating through the boat.

Meanwhile, what of the enemy? I had half expected a shell as soon as we came up, and it was with great anxiety that I looked round. We had been slightly favoured by fortune in that the only thing in sight was a trawler away on the port beam. It was our hunter.

Torplexed

I trimmed right down, hoping to avoid being seen, as it was essential to stay on the surface and get some amperes into the battery. I also altered course away from him.

It was about 5 p.m. that I saw two trawlers ahead, one on each bow. By this time the boat's crew had quite recovered, but I did not wish to dive, as the battery was still pitiably low. I gradually altered course to north-east, but after half an hour's run I almost ran on top of a group of patrols in the dusk. I crash-dived, and they must have seen me go down, as a few minutes later the boat was violently shaken by a depth-charge.

We were at twenty metres, still diving at the time. I consulted the chart, but could find no bottoming ground within fifty miles, a distance which was quite beyond my powers.

At 11 p.m. I simply had to come up again and get a charge on the batteries.

From 7 p.m. to 10 p.m., at regular half-hourly intervals, a depth-charge had gone off somewhere within a radius of two miles of me. Needless to say, I was only crawling along at about one knot and altering course frequently. What was so terrible was the patent fact that the patrols in this area had evidently got some device which enabled them to keep in continual touch with me to a certain extent.

These monotonous and regular depth-charges seemed to say: "We know, Oh! U-boat, that we are somewhere near you, and here is a depth-charge just to tell you that we haven't lost you yet."[17]

As an hour had elapsed since the last depth-charge, I felt fairly happy at coming up, and on making the surface I was delighted to find a pitch-black night and a considerable sea. From 10 p.m. to 1 a.m. I actually had three hours of peace, and in this period I managed to cram a considerable amount of stuff into the batteries. The densities were rising nicely and all seemed well, when I did what I now see was a very foolish thing.

I made my 1 a.m. wireless report to Nordreich, in which I requested orders at 3 a.m. and reported my position, together with the fact that I had been badly hunted. In twenty-five minutes they were on me again! I had most idiotically assumed that the English had no directional wireless in these parts. They have. They've got everything that they have ever tried up there; it was concentrated in that infernal Fair Island Channel.

I was only saved by seeing a destroyer coming straight at me, silhouetted against, the low-lying crescent of a new moon. When I dived she was about six hundred metres away. As I have confessed to doing a foolish thing, I give myself the pleasure of recording a cleverer move on my part. I anticipated depth-charge attack as a matter of course, but instead of

---

[17] Karl was quite right; it is evident that he had the misfortune to encounter one of our new hydrophone-hunting groups, just started In the Fair Island Channel. The incident of the depth-charges every half-hour was known as "Tickling up." Probably the patrol only heard faint noises from him.--ETIENNE.

going down to twenty-five metres, I kept her at twelve. The depth-charges came all right, seven smashing explosions, but, as I had calculated, they were set to go off at about thirty metres, and so were well below me. The boat was thrown bodily up by one, and I think the top of the conning tower must have broken surface, but there was little danger of this being seen in the prevailing water conditions.

I have just had to stop recording my experiences of the past forty-eight hours, as the Navigator, who is on watch, sent down a message to say that smoke was in sight.

The next hour was full of anxiety, but by hauling off to port we managed to lose it. I then had a little food, and I will now conclude my account before trying again to get some sleep.

The account continued.

All my hopes of getting up again that night, both for the purpose of charging and of getting the 3 a.m. signal, were doomed to be disappointed, as the hydrophone operator kept on reporting the noise of destroyers overhead. Occasional distant thuds seemed to indicate a never-ending supply of depth-charges, but they were about four or five miles from me. Perhaps some other unfortunate devil was going through the fires of hell.

At daylight on the second day my position was still miserable. The battery was getting low again, the sea had gone down, and when I put my periscope up at 9 a.m. the horizon seemed to be ringed with patrols. I felt as if I was in an invisible net, and though I endeavoured to conceal my apprehension from the crew, I could see from the listless way they went about their duties that they realized that once again we were near the end of our resources.

All the forenoon we crept along at thirty metres, until the tension was broken at 1 p.m. by a furious depth-charge attack. In some extraordinary way they had located me again and closed in upon me. The first charges were some little distance off, and as they got closer a feeling of desperation overcame me, and I seriously contemplated ending the agony by surfacing and fighting to the last with my gun.

Curiously enough, the procedure that I adopted was the exact opposite.

I decided to dive deep. I went down to 114 metres. At this exceptional depth, three rivets in the pressure hull began to leak, and jets of water with the rigidity of bars of iron shot into the boat. I held on for five minutes, which was sufficient to save me from the depth-charge attack, though two which went off almost above me broke some lamps. I then came up to twenty metres and slowly crawled on. Throughout the long afternoon, though we were not directly attacked again, I heard depth-charges on several occasions sufficiently close to me to demonstrate that these implacable and tireless devils had an idea of the area I was in.

By a supreme effort, working one motor at the only speed it would go, viz., "Dead slow," I managed to squeeze out the battery until I estimated it must be dusk. There was only one thing to do--I surfaced. It was not as dark as I had hoped, and I saw a fairly large sloop-like vessel, about eight thousand metres away, on the port beam. She must have seen me simultaneously, as the flash of a gun darted from her, the shell falling short.

I couldn't dive; there seemed only one thing to do: fight and then die. I ordered the gun's crew up, and the unequal duel began. We were going full speed on the Diesels, and my course was east by north. A good deal of water and spray was flying over the gun, and my crew had little hope of doing much accurate shooting, but I have often found that when one is being fired at there is nothing so comforting as the sound of one's own gun.

Our enemy was armed with two large guns, fifteen centimetres or over, but had no speed, a discovery which raised my hopes again. It was soon evident that, provided we were not heading for another patrol, if we could survive ten minutes' shelling, we should be saved for the time being by the fading light, which was evidently causing our enemy increasing difficulties, as his shots alternated between very short and very much over.

I was actually congratulating the Navigator on our escape, and I had just told the gun's crew to cease firing at the blurred outlines on the port quarter from which the random shells still came, when there was a sheet of yellow flame and a jar which threw me against the signalman. The latter had been standing near the conning-tower hatch, and unfortunately I knocked him off his balance, and he fell with a thud into the upper conning tower. He had the good fortune to escape with a couple of ribs broken, but when I recovered myself and got to my feet, far worse consequences met my eyes.

By the worst of ill-luck, a shell which must have been fired practically at random had hit the gun just below the port trunnion. The result of the explosion was very severe. Four of the seven men at the gun had been blown overboard, the breech worker was uninjured, though from the way he swayed about it was evident that he was dazed, and I expected to see him fall over the side at any moment. The remaining two men were as dead as horse-flesh.

The material damage was even more serious. The gun had been practically thrown out of its cradle, but in the main the trunnion blocks had held firm, and the whole pedestal had been carried over to starboard. The really terrible effects of this injury were not apparent at first sight, but I soon realized them, for an hour later (we had shaken off the sloop) I saw red flame on the horizon, which plainly indicated flaming at the funnel from some destroyer doubtless looking for us at high speed.

I dived, intending to surface again as soon as possible. With this intention in my head, I did not go below the upper conning tower. We had barely got to ten metres, when loud cries from below and the disquieting noise of rushing water told me that something was wrong. I blew all tanks, surfaced, left the First Lieutenant on watch and went below.

There were five centimetres of water on the battery boards, and I understood at once that we could never dive again. For the pedestal of the gun, in being forced over, had strained the longitudinal seam of the pressure hull, to which it is bolted, and a shower of water had come through as soon as we got under. It might have been hoped that this was enough, but no! our cup was not yet full. Chlorine gas suddenly began to fill the fore-end. The salt water running down into the battery tanks had found acid, and though I ordered quantities of soda to be put down into the tank, it became, and still is at the moment of writing, impossible to move forward of the conning tower without putting on a gas mask and oxygen helmet. So we are helpless, and at the mercy of any little trawler, or even the weather.

We have no gun; we cannot dive. The English must know that they have hit us, and every hour I expect to see the hull of a destroyer climb over the horizon astern. We are fortunate in two respects: in that for the time being the weather seems to promise well, and our Diesels are thoroughly sound. We are ordered to Zeebrugge--I could have wished elsewhere for many reasons, but it does not matter, as I cannot believe we are intended to escape.

I feel I would almost welcome an enemy ship, it would soon be over; but this uncertainty and anxiety drags on for hour after hour--and now I cannot sleep, though I haven't slept properly for over seventy hours. I am so worn out that my body screams for sleep, but it is denied to me, and so, lest I go mad, I write; it is better to do this, though my eyes ache and the letters seem to wriggle, than to stand up on the bridge looking for the smoke of our enemies, or to lie in my bunk and count the revolutions of the Diesels; thousands of thousands of thudding beats, one after the other, relentless hammer strokes.

I have endured much.

NOTE BY ETIENNE

A break occurs in Karl von Schenk's diary at this juncture. Fortunately the main outlines of the story are preserved owing to Zoe's long letter, which was in a small packet inside the cover of the second notebook. Zoe's letter will be reproduced in this book in its proper chronological position, but in order to save the reader the trouble of reading the book from the letter back to this point, a brief summary of what took place is given here. The entries in his diary which follow the words "I have endured much," are very meagre for a period which seems to have been about a month in length. There is no further mention of the latter stages of Karl's passage in the wrecked boat to Zeebrugge, so it is presumed that he made that port without further adventure. He was evidently on the verge of a nervous breakdown, and appears to have been suffering from very severe insomnia. He had been hunted for two days, during which he was perpetually on the verge of destruction, and the cumulative effect of such an experience is bound to leave its mark on the strongest man. When he got back to Zeebrugge he must have been at the end of his tether, and whether by chance or design it was when Karl was, as he would have said, "at a low mental ebb" that Zoe made her last and successful attack upon his resolution not to see her again unless she consented to marry him. It is plain from her letter that when he left her after the stormy interview in which he vowed never to see her again, Zoe did not lose hope. She seems to have kept herself au courant with his movements, and actually to have known when he was expected in.

We know that she had many friends amongst the officers, and it is probable that from one of these she was able to get information about Karl's movements.

Bruges was probably a hot-bed of U-boat gossip, and, not unlike the conditions at certain other Naval ports during the war, the ladies were often too well informed. At any rate it appears that Zoe rushed to see Karl directly he arrived at Bruges, and found him a mental and physical wreck, suffering from acute insomnia.

With the impetuous vigour which evidently guided most of her actions, she took complete charge of Karl, and, as he was due for four days' leave, she whisked him off to the forest.

Karl may have protested, but was probably in no state to wish to do so. At her shooting-box in the forest Zoe achieved her desire, and the stubborn struggle between the lovers ended in victory for the woman. There is an entry in Karl's diary which may refer to this period; he simply says, "Slept at last! Oh, what a joy!"

If this entry was written in the forest, it seemed as if Karl had been unable to sleep until Zoe carried him off to the forest peace of her shooting-box and surrounded him with the atmosphere of her tender sympathy.

There is no evidence of the light in which Karl viewed his defeat, when, having regained his strength, he was able to take stock of the changed situation. It is reasonable to suppose that his silence upon this matter in the pages of his diary is evidence that he was ashamed of what he must have considered a great act of weakness on his part.

At all events he realized that he had crossed the Rubicon and that he had better acquiesce in the fait accompli.

He seems to have been in harbour for about six weeks, during which he lived with Zoe, and the lovers enjoyed a brief spell of happiness before Karl set out on his next trip.

Karl seems to have found those six weeks very pleasant ones, though his diary merely contains brief references, such as: "A. day in the country with Z."; "Z. and I went to the Cavalry dance," and other trivial entries--of his thoughts there is not a word.

About the end of 1917 Karl's boat was repaired, and he left for the Atlantic; and once more resumed full entries in his diary.

ETIENNE.

Karl's Diary resumed

Sailed at 9 p.m. last night, and we are now seventeen miles off Beachy Head. The Straits of Dover were frightful; the glare of the acetylene flares on the barrage showed for miles. Seen from a distance it gave me the impression of the gates of hell, through which we had to pass.

I dived, ten miles away, and went through with the tide at a depth of forty metres. Two hours and three quarters of suspense, and at dawn we came up, having passed safely through the great deathtrap. At the moment there is nothing in sight, except a little smoke on the horizon. I am going to dive again till dusk.

2 a.m.

We are thrashing down the Channel with a south-westerly wind right ahead. My instructions are to work for two days between the Lizard and Kinsale Head, and then proceed far out in the Atlantic, where the convoys are supposed to meet the destroyers.

That Fair Island Channel experience was enough for a lifetime. Death, quick, short and sudden, this I am ready for. But torture, slow, long and drawn-out, is not in the bargain which in this year of grace every civilized man and half the savages of the world seem to have had to make with the god Mars.

As I sit in this steel, cigar-shaped mass of machinery, the question rings incessantly in my ears: "To what object is all this war directed, when analysed from the point of view of the individual?" It does not satisfy any longing of mine. I have not got a lust for battle: no one who fights has a lust for battle. Editors of newspapers and people on General Staffs, possibly also Cabinet Ministers, have lusts for battles, as long as they arrange the battle and talk about it afterwards--curse them!

The only thing I want is to be with Zoe. I want to live and spend long years with her, enjoying life--this life of which I have spent half already, and now perhaps it will be taken from me by some other man: some Englishman who doesn't really want to take my life, reckoned as an individual. Around me in the darkness are the patrol boats, manned by the Englishmen who are seeking my life. Seeking it, not to gratify their private emotions, but because we are all in the whirlpool of War and cannot escape.

Like an avalanche, it seems to gather strength and speed as it rolls on, this War of Nations. The world must be mad! I cannot see how it can ever stop. England will never be defeated at sea. We shall conquer on land--then what?

An inconclusive peace.

Even if we smash this island Empire and gain the dominion of the world, how will it advantage me? I can see no way in which I can gain. It would be said, if any one should read this: Gott! what a selfish point of view--he thinks only of his personal gain, not of his country. But, confound it all, I reply, answer me this: Do I exist for my country, or does my country exist for me? For example, does man live for the sake of the Church, or was the Church created for man? Does not my country exist for my benefit? Surely it is so.

Then again, I am risking my all, my life; I live in danger, apprehension and great discomfort; I do all these things, and yet if as a reasonable man I ponder what advantage I am to gain from all these sacrifices I am adjudged selfish. It is all madness; I cannot fathom the meaning of these things.

In position on the Bristol line of approach, the weather is bad.

At twenty metres.

Once again Death has stretched forth his bony fingers to catch me by the throat, and only by a chance have I wriggled free.

Yesterday afternoon at 5 p.m. we sighted a small steamer flying Spanish colours and steering for Cardiff. The weather was choppy, but not too bad, and I decided to exercise the gun's crew, though I did not think there would be much doing, as the Spaniards soon give in. I opened fire at six thousand metres, and pitched a shell ahead of her and ran up the signal to heave-to. The wretched little craft paid no attention, and continued on her lumbering course. I suspected the presence of an Englishman on her bridge, and determined to hit. This we did with our sixth shot, and she stopped dead and wallowed in the trough, with clouds of steam pouring out of her engine-room; we had evidently got the engine-room.

As we closed her, it was evident that a tremendous panic was taking place on board. The port sea boat was being launched, but one fall broke and the occupants fell into the water. My Navigator begged me to give her another, which I did, and hit her right aft. Two boatloads of gesticulating individuals now appeared from the shelter of her lee side and began pulling wildly away from the ship. The Navigator, whose eyes were dancing with excitement, was very keen to play with them by spraying the water with machine-gun bullets; but it seemed to me to be waste of ammunition, and I would not permit it.

Meanwhile we had approached to within about four hundred metres of her port bow. I was debating whether to accelerate her sinking, when I noticed that a fire had broken out aft, and I became possessed with a childish curiosity to see the fire being put out as she sank. It was a kind of contest between the elements.

As I watched her, I was startled to hear three or four reports from the region of the fire.

"Ammunition!" shouted the pilot, with wide-opened eyes.

In an instant I pressed the diving alarm as I realized our deadly peril. Fool that I had been, she was a decoy-ship. They must have realized on board that I had seen through their disguise, for as we began to move forward, under the motors, a trap-door near her bows fell down, the white ensign was broken at the fore, and a 4-inch gun opened fire from the embrasure that was revealed on her side.

We were fortunate in that our conning tower was already right ahead of the enemy, and as I dropped down into the conning tower, I saw that as she could not turn we were safe. A few shells plunged harmlessly into the water near our stern, and then we were under.

We came up to a periscope depth, and I surveyed her from a position off her stern. She was sinking fast, but I felt so furious at being nearly trapped that I could not resist giving her a torpedo; detonation was complete, and a mass of wreckage shot into the air as the hull of the ship disappeared. As to the two boats, I left them to make the best course to land that they could.

As they were fifty miles off the shore when I left them and it blew force six a few hours afterwards, I rather think they have joined the list of "Missing." We are now steering due west to our second position.

Received orders last night to return to base forthwith on the north about route.[18] I have shaped course to pass fifty miles north of Muckle Flugga; no more Fair Island Channel for me.

Statlandlet in sight, with the Norwegian coast looking very lovely under

---

[18] This means into the North Sea round Scotland.—ETIENNE.

the snow--we never saw a ship from north of the Shetlands to this place, when we saw a light cruiser of the Town class steaming south-west at high speed. She had probably been on patrol off this place, where the Inner and Outer Leads join up and ships have to leave the three-mile limit. She was well away from me, and an attack would have been useless. I did not shed any tears; I have lost much of the fire-eating ideas which filled my mind when I first joined this service.

We are due off the mole at 8 p.m. tonight, and my heart leaps with joy at the thought of seeing my Zoe; already I can almost imagine her lovely arms round my neck, her face raised to mine, and all the other wonderful things that make her so glorious in my eyes.

NOTE BY ETIENNE

Before quoting the next entry in Karl's journal it is necessary to explain the situation which confronted him when he arrived in Zeebrugge. In his absence, his beloved Zoe had been arrested as an Allied Agent, and she was tried for espionage within a day or two of his arrival. There is no record of how he heard the news, and the blow he sustained was probably so terrible that whilst there was yet hope he felt no desire to write; but, as will be seen, there came a time when he turned to his journal as the last friend that remained to him. It is a curious fact that, with the exception of an entry at the beginning of this journal, Karl makes little mention of his mother and home at Frankfurt. Though he does not say so, it seems possible that his mother had heard of his entanglement with Zoe, and a barrier had risen between them; this suggestion gains strength from the fact that in his blackest moments of despair he never seems to consider the question of turning to Frankfurt for sympathy. Interest is naturally aroused as to the details of Zoe's trial. The available material consists solely of the long letter she wrote to him from Bruges jail. It may be that one day the German archives of the period of occupation will reveal further details. Information on the subject is possibly at the disposal of the British Intelligence Service, but this would be kept secret. All we know on the matter is derived from the letter, which has been preserved inside the second volume of Karl's diary.

There seems no doubt that she was caught red-handed, but to say more would be to anticipate her own words.

It was a matter of some difficulty to know where best to introduce Zoe's letter, but with a view to securing as much continuity of thought in the story as possible it has been decided to quote it at this juncture, although he did not receive it until after he had made the entry in the journal which will be quoted directly after the letter.

I would like to appeal to any reader who may happen to be engaged in administrative or reconstructive work in Belgium, to communicate with me, care of Messrs. Hutchinson, should he handle any papers dealing with Zoe's trial.

ETIENNE.

Zoe's letter

MY BEST BELOVED,

When you get this letter cease to sorrow for what will have happened, for I shall be at rest, and in peace at last, freed from a world in which I have known bitter sorrow and, until you came into my life, but little joy.

For these past months I am grateful to God, if such a being exists and regulates the conduct of a world gone mad.

For in a few hours I am to die.

It is harder for you than for me; one moment of agony I suffered, a moment that seemed to last a century, when, amidst the sea of faces that swam in a confused mass before me at the trial, I saw your eyes and the torture that you were suffering. When I saw your eyes I knew that the President had said I must die. I am glad that I was told this by you, the only one amongst all these men who loved me. I suppose the President spoke; I never heard him, but I saw your eyes and I knew.

My darling, it was cruel of you to come, cruel to me and cruel to yourself, but I loved you for being there; it showed me that up till the last you would stand by me, and until you read this you cannot know all the facts. That to you, as to the others, I must have seemed a woman spy and that nevertheless you stood by me, is to me a recollection of unsurpassable sweetness, compared with which all other thoughts of you fade into insignificance.

Know now, oh, well beloved, that I was not unworthy of your love.

I have a story to tell you, and I have such a little time left that I must write quickly. The priest who has been with me comes again an hour before the dawn, and he has promised to deliver these my last words of love into your hands.

My real name is Zoe Xenia Olga Sbeiliez, and I was born twenty-nine years ago at my father's country house at Inkovano, near Koniesfol. I am Polish; at least, my father was, and my mother comes from the Don country. There was a day when my father's ancestors were Princes in Poland. Poor Poland was torn by the vultures of Europe, just as your countrymen, my Karl, are tearing poor Belgium and France, and so my family lost estates year by year, and my grandfather is buried somewhere in the dreary steppes of Siberia because he dared to be a Polish patriot.

My father bowed before the storm, and under my mother's influence he never became mixed up with politics. Thus he lived on his estates at Inkovano, and nursed them for my younger brother, Alexandrovitch, the child of his old age. Alex would be nineteen now, had he lived. The estates were large as these things go in Western Europe, but they were but a garden as compared with the lands held by my great-grandfather, Boris Sbeiliez.

My father had a dream, and he dreamed this dream from the day Alex was born to the day they both died in each other's arms.

My father dreamt that one day the Tsars would soften their heart to Poland, and raise her up from the dust to a place amongst the nations, and my father dreamt that Alexandrovitch Sbeiliez would become a leader of Poland, as his ancestors had been before him. And so my father nursed his estates and pinched and saved, in preparation for the day when his beautiful dream should come true.

My poor idealistic father never realized, oh, my Karl, that when one wants a thing one must fight--to the death. Alex was the apple of his eye, but I was much loved by my mother; perhaps she dreamed a dream about me--I know not, but she determined that I should have all that was necessary. Paris, Berlin, Munich, Dresden, and a season in London, then I came home at twenty-one, perfectly educated according to the world, beautiful according to men, and dressed according to Paris. But I was only to find out how little I knew. My mother and I used to take a house in Warsaw for the season, and I met many notable men and women. In these days I, also, thought I could do something for Poland, but after two or three seasons I found that I, too, was only dreaming idle dreams. Oh! my beloved, beware of dreaming idle dreams.

Listen! I once met the Prime Minister of all Russia at a reception. I captivated him, and thought, now! now! I shall do something.

I sat next to him at dinner; I talked of Poland--and I knew my subject--I talked brilliantly; he listened, he hung on my words, and he, the Prime Minister of all Russia, the Tsar's right-hand man, asked me to drive with him next day in his sledge. I, an almost unknown Polish girl!

When I accepted, I was in the seventh heaven of delight.

Next day he called and we set forth; at a deserted spot in the woods near Warsaw he tried to kiss me--I struck him in the face with the butt of his own whip.

That was why he had hung on my words, that was why he had taken me for my drive; it was my Polish body that interested him--not Poland.

The Prime Minister of Russia was confined to his room for two days, "owing to an indisposition." How I laughed when I saw the bulletin in the paper, signed by two doctors, but it taught me a lesson; I never dreamt idle dreams again.

No, I am wrong, my beloved. I dreamt an idle dream, a lovely dream about you and I. An after-the-war dream, if this war should ever end, but like other dreams it has ended--in dreams.

But I must hurry, for my little watch tells me that one hour of my five has gone, and I have much to say.

I could have married, and married brilliantly, but Poland held me back. I did not know what I could do for my country, it all seemed so hopeless, and yet I felt that perhaps one day ... and I felt I ought to be single when that day came.

It was not easy, my Karl, sometimes it was hard; one man there was, Sergius was his Christian name; he loved me madly, and sometimes I thought--but no matter, he is dead now, killed at Tannenberg, and I--well, I will tell you more of my story.

When the war broke out and clouded over that last beautiful summer in 1914 (I wonder will there ever be another like it in your lifetime, my Karl? No, I don't think it can ever be quite the same after all this!), we were all in the country. Alex was back from his school in Petrograd, and my father kept him at home for the autumn term.

How well I remember the excitement, the mobilization, the blessing of the colours, the wave of patriotism which swept over the country; even I, under the influence of the specious proclamations that were issued broadcast by the Government, with their promises of reform, and redress for Poland after the war was over, felt more Russian than Polish. Lies! Lies! Lies! that was what the Government promises were, my Karl.

Under the stress of war the rottenness of that great whited sepulchre, Russia, feared the revival of the Polish spirit; it might have been awkward, and so they lied with their tongues in their cheeks, and we simple Poles believed them; the peasantry flocked to their depots, little knowing whom they fought, but the proclamations which were read to them told them they fought for Poland, and we women worked and prayed for the success of Russian arms.

Then the tide of war swept westward, and all day long and every day the troops, and the guns and the motor-cars and the wagons rolled through the village to the west.

Guarded hints in the papers seemed to say that all was not well in

France, but France was so far away, and all the time the Russians were going west through our village. Mighty Russia was putting forth her strength, and the Austrian debacle was in full swing; these were great days, my Karl, for a Russian!

Then one day the long columns of men and all the traffic seemed to hesitate in the sluggish westward flow, and then it stopped, and then it began to go east. The weeks went on, and one day, very, very faintly, there was a rumbling like a distant thunderstorm. It was the guns! The front was coming back.

Have you ever seen forest fires, my Karl? We had them every autumn in our woods. If you have, then you know how all the small animals and the birds, the rabbits and the foxes, and perhaps a wolf or two, and the deer, and the thrushes and the linnets come out from the shelter of the trees, fleeing blindly from the great peril, anxious only to save their lives. So it was when the front came back. Herds of moujiks, the old men, the women, the children, the poor little babies, struggled blindly eastwards through the village.

Pushing their miserable household gods on handcarts, or staggering along with loads on their backs, and weary children dragging at their arms, the human tide flowed eastwards, round our house, begged perhaps a drink of water, and then wandered feverishly onwards.

They knew not in ninety-nine cases out of a hundred where they were going; their only destination was summed up in the words, "Away from the Front"--away from the ominous rumbling which began to get louder, away from that western horizon which was beginning to have a lurid glow at nights, like a sunset prolonged to dawn.

Then, as the Germans advanced more and more, the character of the tide changed, the civilian element was outnumbered by the military. Companies, battalions, brigades, sometimes in good order, sometimes in no order, marched through the village. They would often halt for a short time, and the officers would come up to the house, where my mother and I gave them what we could. My father lived amongst his books and accounts, and bemoaned the extravagance of the war. Then there were the deserters, the stragglers, the walking wounded, the--but you know, my Karl, what an army in retreat means.

I must proceed with my story, for time moves relentlessly on.

One day a desperately wounded officer, a young Lieutenant of the Guard, a boy of twenty-five, was taken out of a motor ambulance to die.

The ambulance had stopped opposite our gates, and lying on his stretcher he had seen our garden, my garden. He knew he was to die, and he had begged with tears in his eyes to the doctor that he might be left in the garden.

Who could refuse him?

He died within two hours, amongst our flowers, with Alex and I at his side.

Before he died, he begged us, implored us, almost ordered us, to move east before it was too late.

We repeated his arguments to my father, but the latter was obdurate, and he swore that a regiment of angels would not move him from his ancestral home. So we made up our minds to stay.

Things got worse and worse, and one day shells fell in the grounds and we hid in the cellars. That night all our servants ran away, and my father cursed them for cowards. Next day in the early morning we heard machine guns fire outside the village, and then all was still.

At six o'clock Alex, white-faced, came running into the house. He had been down to the gates and he had seen the enemy. They were drunk, he said, and going down the street firing the houses and shooting the people as they came out.

It seemed impossible and yet it was true. It was growing dark, when we heard shouts and saw lights, and from the top of the house I saw a crowd of singing and shouting soldiers, with pine torches, half running, half walking up the drive.

They massed in a body opposite the house. Paralysed with terror, I looked down on the scene, and shuddered to see that every second man seemed to have a bottle. One of them fired a shot at the house, and next I remember a flood of light on the drive, and, in the circle of light, my father standing with hand raised. What my father intended can never be known, for, as he paused and faced the mob, a solitary shot rang out, and he fell in a huddled heap.

As he fell, a boyish voice from the door shouted "Murderers!" It was Alex. With his little pistol I had given him for a birthday present in his hand, he ran forward and, standing over my father's body, head thrown back, he pointed his pistol at the mob and fired twice. A man dropped, there was a flash of steel, the crowd surged forward, and--and, oh! my Karl, they had murdered my beloved brother, my darling Alex.

The next moment they were in the house. I escaped from my window on to the roof of the dairy, and from there down a water-pipe, across the yard to an old hay-loft. For a long time they ran in and out of the house, like ants, looting and pillaging; then there was a great shout, and for some time not a soul came out of the house. I guessed they had got into the cellars. At about midnight I saw that the house was on fire. In a few minutes it was an inferno and the drunken soldiers came pouring out, firing their rifles in all directions.

I had found a piece of rope in the loft. One end I placed on a hook and the other round my neck. I was close to the upper doors of the loft, with a drop to the courtyard, and thus I stayed, for I feared that some soldier, more sober than the rest, might explore the outhouses and find me. I was watching this unearthly spectacle, and never, my best beloved, did I conceive that man could become lower than the beasts, but before my eyes it was so, when I noticed that the great gates at the southern end of the courtyard were opening. As they opened I saw that beyond them were drawn up a line of men. An officer gave an order, and two machine guns were placed in position in the gate entrance; round the guns lay their crews, and the seething mass of revellers saw nothing. I felt that a fearful tragedy was impending, and as I held my breath with anxiety the officer gave a short, sharp movement with his hand and a hideous rattle rose above all noises. The pandemonium that ensued was indescribable. Some ran helplessly into the burning house, others ran round and round in circles, others tried to get into the dairy; one man got upon its roof and fell back dead as soon as his head appeared above the outer wall. The place was surrounded. It was horrible. A few tried to rush for the gate, they melted away like snow before the sun, as their bodies met the pitiless stream of bullets. I suppose two hundred men were killed in as many seconds. The machine guns ceased fire. Ambulance parties came into the yard, collected the dead and living, and within half an hour there was not a soul save myself in the place. Discipline had received its oblation of men's lives.

As an example, it was one of the most wonderful things I have ever known in your wonderful army, my Karl, but it was terrible--terribly cruel.

I never knew what became of my mother, though I feel she is dead-- murdered, perhaps, like my father and my darling Alex, or perhaps she hid somewhere in the house and remained petrified with terror till the flames came. Next morning I left my hiding-place and walked about. Not a German was to be seen, but in the wood was a huge newly-made grave. It was all open warfare then, and this flying column, which was miles in advance of the main body, had moved on. The house was a smoking mass of ruins, but the farm buildings had been spared, and I let out all the poor animals and turned them into the woods, so that they might have their chance.

All day I searched for my father and brother, but not a sign was to be seen, and at dusk I stood alone, faint and broken, amongst the ruins of my ancestors' home. As I looked at this scene of desolation and I contrasted what had been my life twenty-four hours before and what it was then, something seemed to snap in my brain, and for the first time I cried. Oh! the blessed relief of those tears, my Karl, for I was a poor weak, helpless girl, and alone with death and bitterness all round me. Late that night I hid once more in my hay-loft and next morning I left Inkovano for ever. Before I left, I made a vow. It is because of this vow, my beloved, that I am to die. For I vowed by the body of our Saviour and the murdered bodies of my family that, whilst life was in me and the war was maintained, for so long would I work unceasingly for the Allies against Germany. As the war ran its fiery course, I have seen more and more that the Allies are the only ones who will do anything for Poland, my beloved country, so have I been strengthened in my vow.

I struck south on my feet, as a poor girl--I, the daughter of a princely family of Poland! No hardships were too great for me, provided I could reach Allied territory. I travelled from village to village as a singing girl, and once I was driven away with stones by villagers set upon me by a fanatical priest. I came by Cracow, and across the Carpathians, helped to pass the lines by a Hungarian Lieutenant--but I tricked him of his reward; I was not ready for that sacrifice. Then across the Hungarian plains to Buda-Pesth, where I remained three weeks, singing in a third-rate café, to make some money for my next stage. But I had to leave too soon--the old story!--this time it was the proprietor's son. What beasts men are, my Karl! And yet to me you are above all other men, a prince amongst your fellows, and never did I love you so distractedly as that first night at the shooting-box, when I read the scorn in your eyes as you rejected me. I have no shame in telling you this. Am I not already in the grave? And then I must be silent and can only await your coming. After many struggles, wearisome to relate, I came to Hermanstadt, and there, whilst pushing my trade as a dancer, came into touch with a Hungarian band of smugglers, working across the mountain passes between Eastern Hungary and Roumania. I did certain work for these men, and in return crossed with them one bitter night in a thunderstorm into Roumania. At Bukharest I got a good engagement, and when I had saved a thousand marks, I bought a passport for five hundred, and came to Serbia, then staggering beneath the great Austrian offensive.

Once again I was in the horrors of a retreat, but I escaped, reaching

Valona, and crossed to Brindisi, by the aid of a French officer to whom I told my story and who believed me. His name is Pierre Lemansour, and he lives at Bordeaux.

If fortune places him in your power, be kind to him, my Karl, for your Zoe's sake.

I came to Rome; and thence to Paris. I stayed here three weeks, singing in a cabaret. Whilst here I tried to advance my plans in vain! What could I, a poor girl, do for the Allies? The Embassy laughed at me, all except one young attaché who tried to make love to me.

Then I thought of England--England, and her cold, hard islanders, phlegmatic in movements, slow to hate, slow to move, but once roused--ah! they never let go, these islanders!

One of their poets has said: "The mills of God grind slowly, but they grind exceeding small."

That, my Karl, is like England.

They are your most terrible enemies, and you know it.

Do not be angry with me when you read this.

For me it is Poland, for you Germany.

Where I am going in a few hours there is no Poland, no Germany, no England, no war. And perhaps, perhaps, no love.

You and I, Karl, have loved, too well, perchance, but our love was above even the love of countries.

God made the love of men and women, then men and women created their countries.

I see the future before me, Karl, and I foresee that the struggle will be at the end of all things, between England and Germany. One will be in the dust.

Thus, I crossed to England and was swallowed up in the great city of London. England has always had a corner of her calculating heart for the small nations, and in London there is a Polish organization. I applied there, and one day I was taken to the Foreign Office, and found myself alone with a great Englishman. His name was--No, I promised, and it will not matter to you, for though he gave me my chance, I have no love for him, and he will never be in your power. Even as I write these words, he has probably taken a list from a locked safe and neatly ruled a red line through the name Zoe Sbeiliez. I tell you they know everything, these Englishmen. I told him my story, and then he asked me whether I was prepared to do all things for the Allies. I told him I was. He then said that I could go as agent for a back area in Belgium, and my centre would be Bruges. I agreed, and asked him innocently enough how I was to live in Bruges. He looked up from his desk and said:

"You will be given facilities to cross the Belgium-Holland frontier, as a German singer."

"And then?" I asked.

"You will go to Bruges and make friends with an Army officer; he must be high up on the staff."

I guessed what he meant, but hoped against hope, and I said: "How?"

I can still see his fish-like face, hair brushed back with scrupulous care, as without a shadow of emotion he looked up, puffed his pipe, and said in matter-of-fact tones:

"You have a pretty face and an excellent figure. Need I say more?"

I could have struck him in the face. I was speechless, my mind a whirl of conflicting emotions. I was roused by the level tones again.

"Is it too much--for Poland?"

Oh! the cunning of the man; he knew my weakness. Mechanically, I agreed. Certain details were settled, and he pressed a bell. Within five minutes I was walking back to my lodgings.

Thanks to a marvellous organization, which your police will never discover, my Karl, within three weeks I was singing on the Bruges music-hall stage, and accepted without question as being what I was not, a German artist from Dantzig. The men were soon round me, but I had no use for youngsters with money. I wanted a man with information. At last I found my man--the Colonel. He was on the Headquarters staff of the XIth Army, the army of occupation in Belgium, when I first met him. Subsequently he went back to regimental work; but by the time he was killed (and to realize what a release that meant for me, you would have had to have lived with him) I had established regular sources of information concerning which I will say no more. Let your country's agents find them if they can. This must I say for the Colonel: he was a brute and a drunkard, but in his own gross way he loved me, and he licked my boots at my desire, but I had to pay the price. You are a man, and with all your loving sympathy you can but dimly realize what this costs a woman. To me it was a dual sacrifice of honour and life, but it was for Poland, and the memories of my parents and Alex steeled me and strengthened my resolution, and so, and so, my Karl, I paid the price.

My special work was on the military side, and consisted in making quarterly reports on the general dispositions of large bodies of troops, the massing of corps for spring offensives, and big pushes and hammer blows.

Then you came into my life! When the Colonel used to go away it was my habit to mix in the demi-mondaine society of Bruges, to try and live a few hours in which I could forget--oh! don't think the worst! That sort of thing had no attraction for me. I didn't seek oblivion in that direction! I had never even kissed anyone in Bruges until I kissed you that first night we met at dinner--I was attracted to you from the very first; the Colonel was due back in a few days, and I suddenly felt mad, and kissed you. I suppose you put me down as one of the usual kind, out to sell myself at a price varying between a good dinner and the rent of a flat! You will now know that I had already mortgaged my body to Poland.

Then a few days later you will remember we went down for that wonderful day in the forest, and for the first time, Karl, I began to see that I was really caring for you, and a faint realization of the dangers and impossibilities towards which we were drifting crossed my mind.

Do you remember how silent I was on the drive back? In a fashion, my Karl, I could foresee dimly a little of what was going to happen. I had a presentiment that the end would be disaster, but I thrust the idea away from me. Then came the day, just before one of your trips--oh! the agony, my darling, of those days, each an age in length, when you were at sea--when you told me at the flat that you loved me.

How I longed to throw my arms round your neck and abandon myself to your embraces, but I was still strong enough in those days to hold back for both our sakes.

Each time we were together I loved you more and more, and each time when you had gone I seemed to see with clearer vision the fatal and inevitable ending.

But I refused to give up the first real happiness that had been mine in my short and stormy life, and so I clung desperately to my idle dream.

I prayed, I prayed for hours, Karl, that the war might end, for I felt that in this lay our only hope--but what are one woman's prayers, a sinful woman's prayers, to the Creator of all things, and the war ground on in its endless agony just as it does tonight--Karl! Karl! will this torture ever end?

But I must hurry, there is still much to tell you, and Time goes on relentlessly just like the war; it is only life that ends. Then came the days I took you to the shooting-box for the first time, and that night I broke down and, unashamed, offered you myself. Think not too badly of your Zoe, my Karl; when a woman loves as I do, what is convention? A nothing, a straw on the waters of life. I wanted you for my own, passionately and desperately, for I feared that any moment the end might come, and to die without having felt your arms around me would have added a thousand tortures to death. Though I could have welcomed death with joy when I saw the look of sorrowful contempt which you cast upon me that night. Heavens above! but you were strong, my Karl. I am not ugly, and yet you resisted, and I hated and loved you at the same time--oh! I know that sounds impossible, but it isn't for a woman. I slept little that night and, feeling that I could not look you in the face in the morning, I left for Bruges before you got up.

I felt that I could trust you not to try and find out the secret of the shooting-box.

What a relief it is to be able to tell you everything frankly, and how I hated the perpetual game of deception which I had to play.

I used to rack my brains for answers to your perpetual question, "Why won't you marry me?" It was a desperate risk taking you down to the forest, but you loved me so much that you never questioned the reasons I gave you for my secrecy. I can tell you now, Karl, that in the early days when I used to disappear from Bruges, it was to the shooting-box that I went.

But I will write more of that later.

Did you suffer the same agony as I did before you left for Kiel, and your pride would not allow you to come to me? You understand now, my darling, why I could never marry you, and when the Colonel was killed it became harder than ever. Once during that terrible interview before you went up the Russian coast, I nearly gave way and promised to marry you. But how could I? I had sworn my vow, and even tonight, though I stand in the shadow of death, I do not regret my vow.

It is inconceivable that I could have married you and carried on my work--a spy on my husband's country--and if I ever thought of trying to do this impossible thing, a vision which has partially come true always restrained me.

I saw a submarine officer disgraced and perhaps sentenced to death, because his wife had been convicted as a spy!

No! it was impossible.

But if I could not marry you, I still wanted your love.

Then you went up the Russian coast, and I heard of your return in a submarine terribly wrecked. I guessed what you must have gone through, and determined to see you, but when I entered your room and saw you lying open-eyed on your bed, with no one but a clumsy soldier to nurse you, I could have wept. You know the rest; you can perhaps hardly remember how I led you to my car and took you down to the forest. Oh, Karl, are you angry with me for what happened? Do you sometimes think that I took an unfair advantage of your weakness? Please! Please forgive me, you were so helpless, and I loved you so.

Then came those unforgettable weeks whilst your boat was being repaired, weeks which opened to me the door of the paradise I was never to enter. Oh! Karl, I pray that all those memories may remain sweet and unclouded all your life. Think of those days when you think of your Zoe. Alas! they came to an end too soon, and you left for the Atlantic. When you came back all was over; I had been caught at last.

The evidence at the trial was clear enough. I have no complaints. I was fairly caught. You remember the big open space in front of the shooting-box? I do not mind saying now that five times have I been taken up from there in an English aeroplane, and landed there again after two days. Each time I took over a full report on military affairs. Not a word of naval news, my Karl; you will remember I never tried to find out U-boat information. I even warned you to be cautious. Well, they caught me as I landed; the English boy who had flown me back tried hard to save me, but it only cost him his own life.

My first thought was of you, and there is not a jot of evidence against you, save only your friendship for me. Remember this fact, if they persecute you. Admit nothing, believe nothing they tell you, deny everything; they have no evidence; but they are certain to try and trap you.

It was noble of you, Karl, to engage Monsieur Labordin in my defence, but it was useless and may do you harm.

I also know of your efforts with the Governor. I hoped nothing from him, but what you did has made me ready to die; I tremble lest you are compromised.

If only I could feel absolutely certain that I have not dragged you down in my ruin I should face the rifles with a smile.

For my sake be careful, Karl.

When it is all over, cause a few little flowers to cover my resting-place, if this is permitted for a spy. Order them, do not place them yourself; you must not be compromised.

I have told my story, and the end is very near. What else is there to say?

Mere words are empty husks when I try to express my thoughts of you.

Do not sorrow for your Zoe, to whom you have given such happiness.

I am not afraid to die and cross into the unknown, which, however terrible it is, cannot be much worse than this awful war.

Karl! Karl! how I long to kiss you and feel your strong arms crushing the breath from this body of mine which has caused so much sorrow.

Oh, Mother Mary, support me in this hour of trial.

I cannot leave you!

May the Saints guard you and keep you through all the perils of war, and grant that we meet again in the perfect peace of eternity.

Forever,

Your devoted and adoring ZOE.

Karl's Diary resumed

She is dead!

They have killed her, my Zoe, my adorable darling, and I am still alive--under close arrest. Perhaps they will shoot me too, in their insatiable thirst for blood. Oh! if they would! Perhaps, my Zoe, if I could only die and leave this useless world behind, I might find you in the mysterious regions where your spirit now dwells.

Oh! is it well with you, Zoe? Give me a sign--a little sign--that all is well. I have knelt in prayer and asked for a sign, but nothing comes--all is a blank, forbidding and mysterious. Is God angry with us, my Zoe, that we sinned before Him? Surely, surely He understands. He must have mercy on me if He is going to make me go on living. If this is my punishment, I can bear it; I will live without you happily if only I may know that all is well with you.

Your letter, Zoe! Can you read these words as I write; can you sense my thoughts? Speak! Ah! I thought I heard your voice, and it was only the laughter of a woman in the street. Your letter has filled me with joy and sorrow. I read and re-read the wonderful words in which you say you loved me from the beginning, but when you plead that I shall not turn in loathing from your memory--with these words you smash me to the ground. Most glorious woman, I never loved you so well and so passionately as the day you stood at the trial, ringed round with the wolves, the clever lawyers, the stolid witnesses, the ponderous books, the cynical air of religious solemnity with which the machinery of the law thinly cloaks its lust for blood--for a life.

Even when my ears heard the sentence, I could not believe it would be carried out. The firing party, the chair, the bandage. Oh, God! spare me these awful thoughts. To think of your breasts lacerated by the----Oh! this is unendurable! Stop, madman that I am!

⊕   ⊕   ⊕

I am calmer now; I have read your letter again and rescued the journal from the grate into which I flung it. The fire was out; I am not sorry; my journal is all I have left, and in its pages are enshrined small, feeble word-pictures of paradise on earth. To read them is to catch an echo of the music we both loved so well. Music! you were all music to me, my Zoe. Your voice, your movements, your caresses all seemed to me to speak of music.

I ask myself, I shall always ask myself until the last hour, whether all that could be done to save you was done. I tried to telegraph to the Kaiser for you, Zoe, but the wire never got further than Bruges post office; they stopped it, and put me under arrest. It was only open arrest, my darling, and on that last awful night I forced them to let me see the Governor. I, Karl Von Schenk, knelt at his feet and begged for your life. He simply said, "You are mad." I left the Palace under close arrest.

Was ever woman's nobleness of character so exemplified as in your life? Be comforted, Zoe, that in all my black sorrow I cling desperately to my pride in your strength. I long to shout abroad what you did and why you would never marry me, to tell all the gaping world that when you died a martyr to duty was killed. I am so unworthy of what you did for me, my darling, and it tortures me with mental rendings to think that whilst I prided myself in my strength of mind, I was dragging you through the fires of hell. When I think of those six weeks we had together, my brain says, "And they might have been months had you not spurned her in the forest."

Oh, Zoe! if the priests say truth and all things are now revealed to you, forgive me for this act of mine. Come to me in spirit and give me mental peace.

As I write like this, as if it was a letter that you might read, I am comforted a little; I rely utterly on the hope, which I struggle to change into belief, that you can read this and know my thoughts. For when I think that had things been otherwise you might have been leaning over my chair at this moment, and running your cool fingers through my stiff hair; when I think of this, my darling, the full realization comes to me of the gulf which must divide us for some uncertain period, and the lines of this page run mistily before my eyes.

Zoe, my Zoe, strange things have happened in this war; wives declare they have seen their husbands, mothers have felt the presence of their sons; if the powers permit, come to me once again, I implore you, and give me strength to live my life alone.

Examined before the Court of Inquiry today. Fools! can't they realize that I don't care if they do shoot me?

In the Mess, people avoid me. What do I care? Not one of them is worthy to stand on the same soil that holds her beloved body. They have buried her in the Castle grounds. In accordance with her wishes, I have arranged for flowers. Perhaps one day when all this is over I may be able to live here and tend the place where she sleeps, free at last from all her cares.

At the Court of Inquiry they tried to cross-examine me on our life together. Dolts! what do they aim at proving? That I loved you? I hardly listened. When they finished the evidence, the President asked me if I had anything to say! Anything to say! I felt like telling them they were cogs in the most monstrous machine for manufacturing sorrow and destruction that mankind had ever devised. I could have shaken my fist in their solemn faces and shouted "Beasts! you murdered her! You destroyed that most wonderful woman who lowered herself to love me."

Actually there was a long silence, and then the Vice-President, Captain Fruhlingsohn, said, "Speak; we wish you well." It was the first touch of sympathy, the only sign of humanity I had received in all these awful days, and it touched my stubborn heart and the longed-for tears flowed at last.

I murmured: "Gentlemen, I am no traitor; but I loved her as my own soul."

"Dissolve the Court. Remove the prisoner." Like the clash of iron gates, officialdom came into its own again.

⊕ ⊕ ⊕

So I am not to be shot! Not even imprisoned! "Don't fall in love with enemy agents again!"--that summarized their verdict.

Ha! Ha! Ha! It is all horribly funny. The real reason is that they need me. I am a trained and skilful slaughterer on the seas; I am an essential part of the great machine. And they haven't got any spares! I was in the Mess yesterday when the English papers we get from Amsterdam arrived. Oh! a pretty surprise awaited the first man who opened The Times. These English had published the names of 150 U-boat commanders they had caught. There they all were. Christian names and all complete. The only thing missing was a blank space in which to fill in our names when the time comes.

Dinner was a silent meal last night, and next morning some rat of a Belgian had posted the list on the gatepost of the Mess. The machine has offered five hundred marks for his capture--how foolish; as if by shooting him they would take any names off the long list.

I am to sail at dawn tomorrow. I shall not be sorry to get away for a space from this place with its mingled memories of delight and death.

Back again,

...and I haven't written a word for three weeks.

My billet last trip was off Finisterre. I sighted two convoys, but there were destroyers there; they are so black and swift I don't go near them. I don't want to die in a U-boat. It's not worth while. It is easy to avoid these convoys. I dive and make a great fuss of attacking, then I steer divergently. Nobody knows where the enemy is except me; I am the only one who looks through the periscope--I take good care of that. And then how I curse and swear when I announce that the convoy has altered course, and there is no chance of getting in to attack. None of them are so disappointed as I am!

The mines get on my nerves, there is no way of dodging them, and Lord! how they sprout on the Flanders coast. I am to go out in six days. It is very little rest. I believe they want to kill me. But I won't die! Not I.

I went to her grave yesterday for the first time. I had thought I should weep, but I did not; in fact it left me quite unmoved. I feel she's not really dead; she comes to me sometimes, always at night when I am alone and when we are at sea. There's nothing very tangible, but I catch an echo of her voice in the surge of the sea along the casing, or the sound of the breeze as it plays along the aerial. And so I will not die until she calls me, for up to the present her messages have told me to live and endure.

A very awkward incident took place last night. We were off the Naze and saw a steamer some distance away. We dived to attack. When we were about a mile away I had a look at her, and something about her put me off. I half thought she was a decoy ship, and I privately determined I would not attack. I steered a course which brought me well on her quarter, and as soon as I saw that it was impossible to get into position to fire I increased speed on the engines and shook the whole boat in efforts which were ostensibly directed to getting her into position. At length I eased speed and bitterly exclaimed that my luck was out.

The First Lieutenant suggested that we should give her gunfire, but I pointed out that I had good reason to suspect her of being a wolf in sheep's clothing, and as he had not seen her he could hardly question my judgment. I was going forward, when I accidentally overheard the Navigator and the Engineer talking in the wardroom. I listened.

The Engineer said: "The Captain doesn't seem to have the luck he used to command."

"Or else he has lost skill!" replied Ebert. "We never fired a torpedo at all last trip, and it looks as if we are following that precedent this time."

I had heard enough, and, without their realizing my presence, I returned to the control room. I considered the situation, and came to the conclusion that they suspected nothing, but it was evident that their minds were running on lines of thought which might be dangerous. I looked at my watch and saw that there was still two hours of daylight left, and then decided to play a trick on them all. I relieved the First Lieutenant at the periscope, and when a decent interval of about half an hour had elapsed I saw a ship. This vessel of my imagination, a veritable Flying Dutchman in fact, I proceeded to attack, and, after about twenty minutes of frequent alterations of speed and course, I electrified the boat by bringing the bow tubes to the ready.

The usual delay was most artistically arranged, and then I fired. With secret amusement I watched the two expensive weapons of war rushing along, but destined to sink ingloriously in the ocean, instead of burying themselves in the vitals of a ship. An oath from myself and an order to take the boat to twenty metres. With gloomy countenance I curtly remarked: "The port torpedo broke surface and then dived underneath her, the starboard one missed astern."

So far all had gone well, but ten minutes later I nearly made a fatal error. We had been diving for several hours, the atmosphere was bad, and as it was dusk I decided to come up, ventilate, and put a charge on the batteries. I gave the necessary orders, and was on my way up the conning tower to open the outer hatch. The coxswain had just announced that the boat was on the surface, when a terrible thought paralysed me, and I clung helplessly to the ladder trying to think out the situation. It had just occurred to me that as soon as the officers and crew came on deck they would naturally look for the steamer we had recently fired at; this ship in the time interval which had elapsed would still be in sight.

As I came down, the First Lieutenant was at the periscope, looking round the horizon. Quickly I thrust the youth from the eyepiece, and, as calmly as I could, said: "I thought I heard propellers."

Half an hour later we surfaced for the night. I have been wondering ever since whether they suspect, for the three of them were talking in the wardroom after dinner and stopped suddenly when I came in.

I must be careful in future.

I was sent for this morning by the Commodore's office, and handed my appointment as Senior Lieutenant at the barracks Wilhelmshafen. No explanation, though I suspected something of the sort was coming, as three days after we got in from my last trip I was examined by the medical board attached to the flotilla.

So I am to leave the U-boat service, and leave it under a cloud! It is a sad come-down from Captain of a U-boat to Lieutenant in barracks, a job reserved for the medically unfit for sea service. Am I sorry? No, I think I am glad. Life here at Bruges is one long painful episode. No one speaks to me in the Mess. I am left severely alone with my memories. The night before last I found a revolver in my room, and attached to it was a piece of paper bearing the words: "From a friend."

Perhaps at Wilhelmshafen it will be different, and yet, when I went down to the boat at noon and collected my personal affairs and stepped over her side for the last time, I could not check a feeling of great sadness. We had endured much together, my boat and I, and the parting was hard.

At Barracks

As I suspected when I was appointed here, my job is deadly to a degree, and my main duty is to sign leave passes.

Our great effort in France has failed, and now the Allies react furiously. The great war machine is strained to its utmost capacity; can it endure the load? Our proper move is to paralyse the Allied offensive by striking with all our naval weight at his cross-channel communications. The U-boat war is too slow, and time is not on our side, whilst a hammer blow down the Channel might do great things. But we have no naval imagination, and who am I, that I should advance an opinion? A discredited Lieutenant in barracks--that's all.

Worse and worse--there are rumours of troubles in the Fleet taking place under certain conditions. It is the beginning of the end!

Last night the High Seas Fleet were ordered to weigh at 8 a.m. this morning. A mutiny broke out in the König and quickly spread. By 9 a.m. half a dozen ships were flying the red flag, and today Wilhelmshafen is being administered by the Council of Soldiers and Sailors. There has been little disorder; the men have been unanimous in declaring that they would not go to sea for a last useless massacre, a last oblation on the bloodstained altars of war. Can they be blamed? Of what use would such sacrifice be? Yet to an officer it is all very sad and disheartening.

I have seen enough to sicken me of the whole German system of making war, and yet if the call came I know I would gladly go forth and die when tout est perdu fors l'honneur. Such instincts are bred deep into the men of families such as mine. We approach the culmination of events. Today Germany has called for an armistice. It has been inevitable since our Allies began falling away from us like rotten print. The terms will doubtless be hard.

Heavens above! but the terms are crushing! All the U-boats to be surrendered, the High Seas Fleet interned; why not say "surrendered" straight out, it will come to that, unless we blow them up in German ports. The end of Kaiserdom has come; we are virtually a republic; it is all like a dream.

We have signed, and the last shot of the world-war has been fired. Here everything is confusion; the saner elements are trying to keep order, the roughs are going round the dockyard and ships, looting freely. "Better we should steal them than the English," and "There is no Government, so all is free," are two of their cries.

There has been a little shooting in the streets, and it is not safe for officers to move about in uniform, though, on the whole, I have experienced little difficulty.

I was summoned today before the Local Council, which is run by a man who was a Petty Officer of signals in the König. He recognized me and looked away. I was instructed to take U.122 over to Harwich for surrender to the English. I made no difficulty; some one has got to do it, and I verily believe I am indifferent to all emotions.

We sail in convoy on the day after tomorrow; that is to say, if the crew condescend to fuel the boat in time. Three looters were executed today in the dockyard and this has had a steadying effect on the worst elements.

I went on board 122 today, and on showing my authority which was signed by the Council (which has now become the Council of Soldiers, Sailors and Workmen), the crew of the boat held a meeting at which I was not invited to be present. At its conclusion the coxswain came up to me and informed me that a resolution had been carried by seventeen votes to ten, to the effect that I was to be obeyed as Captain of the boat. I begged him to convey to the crew my gratification, and expressed the hope that I should give satisfaction. I am afraid the sarcasm was quite lost on them.

We are within sixty miles of Harwich and I expect to sight the English cruisers any moment. I wrote some days ago that I was incapable of any emotion. I was wrong, as I have been so often during the last two years. In fact, I have come to the conclusion that I am no psychologist--I don't believe we Germans are any good at psychology, and that's the root reason why we've failed. I do feel emotion--it's terrible; the shame--the humiliation is unbearable. I wonder how the English will behave? What a day of triumph for them.

The signalman has just come down and reported British cruisers right ahead; it will soon be over. I must go up on deck and exercise my functions as elected Captain of U.122, and representative of Germany in defeat. One last effort is demanded, and then----

NOTE

This is the last sentence in the diary. It is probable that he suddenly had to hurry on deck and in the subsequent affairs forgot to rescue his diary from the locker in which he had thrust it.

ETIENNE

THE END

USS Casimir Pulaski (SSBN-633)
The Story of a Cold War Warrior
Don Murphy

There Were No Bullets Flying Overhead...

...nor the sound of artillery. Indeed, nothing to let you know that we were 'at war'. But make no bones about it; the vast ocean spaces of the Atlantic and Pacific were fertile breeding grounds for violence and a virtual state of war existed between us and our communist naval forces counterparts. Following an SSBN was an act of war. So every effort was made to make sure that it didn't happen. Silence was the name of the game. While planes jockeyed for position in mock dogfights and ships played chicken with each other, the submarine fleet, like the cavalry of wars past, remained in the background, waiting to come to the rescue, bugle's tooting and flags flying in the wind. The odds were stacked in the West's favor. SOSUS and other listening sensors littered the vital sea-lanes through which Soviet submarines had to transit. Even NATO navies with few ships boasted impressive land-based ASW assets. Everyone, save for a few, flew the Sea King helo with its serious sub-hunting capability. And then there were the fast boats. Over one hundred and fifty of Ronald Reagan's '600-ship navy' were submarines. Help would never be far away should we need it. Our 'little brothers' varied but almost always ended up to be *USS Atlanta* or *USS Minneapolis-St. Paul*; serious Los Angeles class killer subs. *USS Atlanta* crept into port every now and then, but *USS Minneapolis-St. Paul* was a mystery. She buzzed us once at high speed like a high school kid screeching down Main Street in his Camaro. *This was no Camaro.* This was several thousand tons of armed to the teeth steel. After the humiliation of being 'caught with our pants down' subsided, silent prayers were offered, thanking any creator who cared to listen that we were indeed grateful that she was on our side! Any joking subsided once the hatch closed. Our job was to silently live in a patrol area and make sure that if the poo-poo hit the fan, we were ready to plug the source with sixteen multiple warhead missiles.

The patrol was our reason for living. The patrol was the reason we commanded the resources we did. The food alone was unreal. The submarine cook spoiled his crew daily. Steak and lobster every other Sunday. Fresh baked breads and cakes. Custom, made to order omelettes. And all in a never-ending supply! The normal sailor dreamed of such a feast. The submarine sailor took it in his stride. "Doesn't everyone have pizza baked to order every Friday night?" Once the hatch closed, any news we had seen or read prior to the hatch coming down was all we had. There was no CNN or Fox News being piped in twenty-four-seven. This made life very interesting. A surface sailor onboard a destroyer or cruiser would see a crisis develop slowly over several days. We, on the other hand, would get an order to launch missiles and then have to do it. Numerous drills and exercises would be run to ensure that we did what we were supposed to do. Indeed, it wasn't known whether the launch order was real or fake until a certain part of the missile launch process was reached. Several of these tests were run every patrol, so no, Denzel Washington's character in *Crimson Tide* would have been weeded out long before he got to be an executive officer. The entire time I was on the *Pulaski*, we never had anyone that I recall refuse to obey a launch order. The Weapon System Readiness Test - or WSRT - was practiced until perfection was achieved. The Ship's Inertial Navigation System (SINS) would update every so many periods and then the missiles would know where they were when launched and where they were headed.

Patrol started, technically, when the R & R period ended. Two months of intense training going over the newest Navy regulations and directives, reviewing lessons learned from our last patrol and those of other submarines. There were missile team trainers and diving trainers which taught new crew how to drive the submarine, service the weapons, etc. The Navy will argue with me, but this training period was for the sole purpose of preparing to go to sea. So technically, this is when patrol 'started'. The initial batch of Ohio class submarines had a nice purpose-built submarine base in Bangor, Washington. The next batch would go to the next purpose-built base and the largest Western submarine base, King's Bay, Georgia. Unfortunately, we were there as well, which meant that the third month of the cycle, refit would take place away from home. So at the end of two months, we'd say goodbye to wives, girlfriends and family and load up on a bus and travel four hours to Georgia. Wives and girlfriends could visit and we'd either have them come down for the weekend or we'd pile into a car and head home for a long weekend. Refit started with the emotional 'bus away' when we left Charleston, South Carolina. To make the whole situation even more unbearable, bus away was done in dress uniform.

At this point in time, *USS Canopus* (AS-34) IS the submarine base. As time goes on, the base will become one of the biggest submarine bases in the world. As sad as the Pulaski's passing into the pages of history is, the plight of the tenders is even more depressing. It seems that their contribution to our nation's defense is not worth so much as a page in a book or a monument.

Repairs of all sorts are affected and maintenance is undertaken on virtually every system. The boat is normally fitted with a scaffold straight away to facilitate antenna/mast work and work on the sail top itself. During Cold War hysteria with stories circulating of Russian 'Spetnatz' special forces teams, the scaffold also serves as a watch tower with anti-swimmer lights and shooting stands for the sail sentry.

All coamings and fairings are frequently and thoroughly inspected to ensure optimum mast performance. Remember, a submarine lives 'blind' and therefore, reliance on periscopes and antennae is mandatory. Cleaning and preservation work will be carried out up until the day patrol starts. The cycle is truly exhausting.

Stores load is truly an amazing adventure. It is astounding to the untrained observer just how much 'stuff' is needed before we can go to sea. How many burgers does our freezer hold? How many reams of paper, pencils and pens do we have? The constant din of the tender's cranes drowns out everything as pallet after pallet of everything from soup to nuts is brought onboard.

A Ship That Deliberately Sinks...

What kind of guy volunteers for service in such a ship? Probably the same kind of guy who'd agree with jumping out of a perfectly good airplane (paratrooper). The *Casimir Pulaski* was divided into two crews: a Blue and a Gold crew, each with their own commanding officer, executive officer, engineer, navigator, etc. The crew was a cross section of Navy occupations with a good chunk of the Navy's trades being represented. Unlike the surface and aviation communities of the U.S. Navy, the submarine force was strictly volunteer. No one was here that did not want to be here. This produced an espirit de corps lacking in a lot of units. And if you needed some spring in your step while working on your prideful swagger, it didn't do any harm to know that your vessel was the most powerful weapon ever built by man. Each Trident I C4 missile carried multiple warheads capable of reducing a target to fine ash in next to no time. And there was capacity to carry sixteen of them. Contrary to popular myth, the doors and hatches were quite generous and it was possible for overly tall or large men to serve onboard boats. Standards were relaxed. Anyone who wanted one could have a moustache as long as it remained within naval regulations. This was unlike the surface Navy and its authoritarianism. Officers were different. We addressed them as 'Mister' and there was mutual respect. You'd lay your life down for any of the officers and vice versa. The captain was one of the most powerful men in the world when he put that key around his neck. He had an open door policy and was the epitome of fairness, yet still in control. The benefits of submarine service were equally lucrative. Unlike sailors in the rest of the Navy, who received sea pay *only* when they were physically at sea, submariners received sea pay all year round. And then on top of that, there was the Family Separation Allowance and Submarine Pay. This equaled several hundred extra dollars per month, which, in 1980s dollars, were a boon.

Then there was the infamous 'FBM Deduction' on your income tax form. One more way the grateful nation paid you back for keeping them safe at night. Yessiree...we were part of the nation's defense. Not just some shmoe military grunts. Local merchants extended free memberships to buying clubs. Disney tickets? Just ask. We didn't drive the local economy; we *were* the local economy. We all lived in the same apartment complexes and after returning from sea, there would usually be a new complex sprouting up nearby. All competed viciously for our dollars. There were clubs and bars all geared towards keeping us happy. From Rivers Avenue to Goose Creek, there was Navy as far as the eye could see. St. Mary's was soon referred to as 'King's Bay', even though King's Bay was the base name. Rules were bent regularly. Only submarine wives and children received free dental care. The rest of the Navy was on their own when it came to their families. 'If we can squeeze you in' was the motto. Not so for the sub sailors. Our wives commanded the same respect. With their bumper stickers proclaiming 'sub sailors do it deeper' and 'my man takes the subway to work', they had the hardest job: not knowing. Even today with sailors having email and payphones onboard ship, the submarine sailor is still separated from all communication with his family. The sole purpose of the boat's radio gear is to receive either an 'all clear' message or a message saying 'it's time to launch'. So the submarine wife/girlfriend doesn't hear from her man until he returns. Submarine wife is the toughest job in the Navy.

A Lifetime of Memories

The sea groans as it does and if one listens closely, as when placing a seashell to one's ear, you can almost hear the voices of ships long past. My trips to the seaside are no different and voices call out as if the sea itself is searching for a homeport for all its lost ships.

We cursed the *Pulaski* - God did we curse her! She stank us from head to toe. Upon return from patrol, loved ones banished us to the shower to soak for hours in an effort to remove the smell her oxygen scrubbers bestowed on us. And her reliability! Damn her reliability! Picture yourself at sea, stretched mentally and emotionally. The end of patrol looms but - your replacement needs repair and can't make it. Extension time. And there was good old 'CP' – 'old faithful' - remaining on station for extra time. She never broke, damn her. And her filth...she took forever to clean! And no matter how hard or how long you cleaned, she took pleasure in displaying spots you missed.

Despite it all, she never changed her sound. The rhythmic humming of her power plant and the noises made by various fans and motors were soothing. Sleep came easy inside one of the old girl's bunks. Day after day she carried us back and forth through hostile waters. As old as she was, her sonar suite never failed to protect us, her reactor never failed to power us.

There was bitter emotion upon seeing her berthed. Why on earth after leaving your family would you be happy to see *her*? It was twisted but she knew our names. And once on board, it was as if we had never left. The knobs were almost in the same position; the spilt soda seemed to be there from last patrol and a bunch of other little details that comprise a halfway decent haunted house story.

She was my first boat and I suppose it burned that I couldn't be there for her passing. One only wonders what she was thinking as her flag was hauled down for the last time, only to be stuffed in a desk drawer somewhere. Who was the last to salute her? An original crewmember, or some nub who was just there when the Navy came to round people up?

We all fantasized about our last trip on her. Was it orders to shore duty? Glorious retirement after years of faithful service? Yard duty? Or perhaps to another boat? My last day crept up quickly. I was awaiting test results for an MRI that I had taken. The phone call came early in the morning and by mid-afternoon I was off the boat. I packed excitedly as one would, knowing that later that evening I'd be in my wife's arms instead of separated. And this was to be a Christmas patrol as well, meaning I'd miss our time together had I gone to sea.

In the past, I envisioned myself making a walk through of her before I left. Shaft alley where I hid from the COB numerous times, AMR 1 where I failed my first walkthrough, and finishing at the torpedo room. Word swept through the boat like wildfire that I wasn't making patrol. As it was a Christmas patrol, I was presumptuous if I expected widespread happiness at my departure! My emotions then centered on the completely narcissistic thought that I wouldn't be getting another star for my patrol pin. Luckily the XO broke my chain of thought by asking if I'd like to drive his car home for him?

The answer was a no-brainer as I had no other way to get home. I arrived home and within a day or two embarked on a whirlwind medical discharge, comprising a medical board in Washington, DC and numerous other interviews and meetings. The Cold War was over so George Bush Sr. had sailors to trim from the fleet. My mistake was being on medical hold at the time. I had letters from admirals requesting I be retained. No such luck. Four months later I stood in the commander's office at Transient Personnel Unit SIX and shook her hand goodbye.

The dreams of *Pulaski* started shortly afterwards almost like some kind of curse or spell. The ball cap that I hated wearing suddenly became a favored piece of apparel. I was still young and thoughts of death - animate and inanimate - were far from me. I always assumed that one day my son would salute her flag and request permission to come aboard. I was wrong.

I doubt if she still had her name when Joe, the shipyard worker, lit his torch and started cutting into her hull. The cutting was no doubt easy, helped along by years of exposure to sea pressure. I often wonder which part went first. More often though, what was made of her high tensile steel? What plowshare was she beaten into? Was her steel used to make hospital beds, or perhaps crutches for children? How about girders for a retirement home? It must have been an important cause, surely.

My son asked out of the blue about her. And one day, with him and his friends in tow, we made the pilgrimage to King's Bay, Georgia. The road grew familiar too quickly. The front gate hadn't changed. Lower base, however, may as well have been another country. There was building and construction everywhere. Not an empty lot stood where there was once swamp and wilderness. The huge garages for the Ohio class boats were not only completed but had boats in them.

We were met by the executive officer of *USS Rhode Island* who gave us a tour. The tour was long and complete. As complete as could be given due to security considerations. The boys were duly impressed, as was I. The 'Hotel' class was the name we'd given to them back in the day. Monsters of the deep with every provision. No hand-to-hand stores load on this boat! A huge elevator to take whole pallets at a time! And not sixteen, but twenty-four missiles. The smell of amine never changed, that sickening chemical used to scrub the oxygen, which permeated everything you wore.

USS Canopus was long gone and her jetty had been swallowed by the massive pier complex. Warrior Wharf stood ominous with its monolith commemorating the loss of *USS Thresher* and *USS Scorpion*. I remember the dedication of that pier as if it was yesterday. *USS Canopus* and the *Pulaski* stood off to the side. I could almost see both of them still there.

There was no memorial to the sacrifices made by the original forty-one SSBN's. Not even a marker by the pier. All who sailed on them get a 'Cold War' certificate, issued in caring government style after numerous tricks have been performed as if by a hound begging for table scraps. A scrap of paper to make up for lifetimes lost.

The Cold War's toll will never fully be realized. On *Pulaski* alone there were families that broke apart due to the strain of sea duty and patrols.

There were men crippled in accidents that occurred during refit or repair or on patrol itself. And it must have been just as bad on the Soviet side.

Leaving the base took a while. We left and arrived at Fort Clinch. As we walked the ramparts my thoughts took me back to *Pulaski* crossing the fort, heading out to sea. Her pilot had left her by then and she was heading out on her own. A friendly Los Angeles and Orion would greet her at the ocean's edge and she'd clear her bridge and gently nose down into the deep. The blowing of her main ballast tanks would create a small plume like some happy whale diving down to its home. As I scanned the Atlantic I thought I saw what resembled a black sail, but it was just my imagination and some wishful thinking.

There is a heaven for submarines and *Pulaski* is there. Her baffle is always clear, her tanks empty, her fresh water supply topped up and her reactor is at full power. The smell of fresh-baked bread fills her passageways. At peace finally, she'll have her number and name proudly emblazoned on her hull. She and her sister boats will all sail together and tell tales of their past glories.

Fair winds and following seas...

SUBSIM ROLL CALL

Not by forum post counts, but by word count, probably nobody else has typed as much in the Subsim forums as me. I joined when Sub Command was still an idea, and I just had found Jane's 688(I). I had a question on it, searched around, stumbled over this place that ironically already was called "Neal's Subsim Review" and posted my question. Being a Steel Beasts fan, I regularly update the tanksim forum, and have spent a lot of time (too much time, many would say) in the general topics forum. I still hold up interest in SH3, while my interest in Sub Command and Dangerous Waters has faded. Subsim is *the* place for me to go to learn about modern and historical naval warfare, new subsim releases, and read the cool articles. Thanks for many years of service and enjoyment, Subsim.

Marc "Skybird" Hoell

My subsimming 'revelation' came when I was going to meet a mate for some beers. He was an engineering rating on *HMS Spartan* and when I called at his cabin in the shore base he showed me a new game he'd bought – Sub Command. I was impressed with the graphics, a step change from 688(I) which I was still playing at the time. What really sold it for me was when he told me he'd gotten advice from a couple of sonar watchkeepers on his sub on how to use the sonar – they laughed at him a bit until they saw the game and the were impressed with the realism. That accolade sold it to me, so I upgraded my old PC and bought the game. I haven't looked back since.

Les 'lesrae' Robertson

As a kid growing up in Los Angeles during the Fifties, I had the opportunity to see the TV series *Silent Service*. It was kind of a documentary about WWII submarines. I loved the opening sequence where the submarine comes barreling up to the surface at flank speed. I would watch the show, then off to bed, only to lay awake and dream of being a sub sailor. I even got to build a model of the *U-505* from Revelle as a teenager. After high school, and with the Vietnam War going full bore, I decided it was time to join the U.S. Navy. I promptly volunteered for Submarine Service. After Sub School and Torpedoman School, I was assigned to the WWII sub *Segundo* SS-398. What a thrill to be part of diesel boat history.

In 2005 I retired from my job. I purchased my first computer and my son bought me my first computer game, Silent Hunter III. I had died and gone to heaven. I learned about Subsim and immediately began reading the forum. I was surprised to see how many younger people were interested in submarines. Subsim has been not only useful for information, but most importantly it's the great members who are here, always helpful and friendly. Thanks, Neal, for bringing all of us together.

Mike "sunvalleyslim" Bissett

Watching documentaries and reading books about the wars and the men who fought them can be quite sobering. We can't put ourselves in their shoes no matter how hard we try. However, not to minimize what those men did, with a bit of imagination and the right tools, we as sub gamers and historians can take a the briefest glance into their worlds. It never ceases to amaze me when I catch myself holding my breath, sitting at the edge of my chair listening to that fading destroyer contact; or the adulation when your manual targeting solution results in an eruption of fire and water. For those of us who love melding history and gaming, it's usually a bittersweet victory, because we know, all too well, what these games were based on, fun as they may be.

Mark Prohaska (jbt308)

My fascination with computers began when my father picked up a Sol-20. The only game we had was a Star Trek game. That machine had a whopping 32k. When the Commodore 64 came out, I thought I was in heaven -- it was in COLOR! I have always loved subs thanks to Ed Beach's *Run Silent, Run Deep*. My first sub sim was Microprose's Silent Service II. Things have come a long way since then, through GATO, Red Storm Rising, Hunt for Red October, Jane's 688(I), Sub Command, and my latest 'sub love', Silent Hunter III. What really amazes me about SH3 though, is the community that has built up to support and enhance it. Most of the enhancements are free. Incredibly talented people giving of their skills and time to make my experience more real and more enjoyable. I thank them all. I also thank Neal Stevens, a man who not only knows how to spell his first name correctly, but lives only a few miles as the crow flies from where this old grandpa grew up.

Neal "NealT" Truitt

I started out as a hard-core fighter aircraft simmer with the release of Microprose's "F-19" in 1988 and was hooked on software that was as realistic as possible. A combination of playing classics like Red Storm Rising and 688 Attack Sub made me a "Nuke" fanatic. By high school I preferred to battle Sierras, Alfas and Victor IIIs under the arctic icepack instead of ninjas in "Mortal Kombat."

Despite my narrow interests, I married my beautiful wife who puts up with my hobbies of cracking keels and model-building. I flew helicopters for a while before becoming a high school science teacher. I gave SHIII a try after mastering Dangerous Waters and today I like conventional boats just as much as nukes.

Matt "Snake-Eyes" Sabin

I joined Subsim this year. Found it by accident while searching for Internet for information on SH4. With my first look, I was excited and just blown away! The talents on Subsim for SH4 are unbelievable. The gameplay experience has increased 20x over the stock level and it's great! I'm an old 688(I) US submarine sailor. I have always missed the service that I left behind, Subsim is my feed to my life long addition. I'm proud to be a Submariner and a Subsim member.

Norm "Chopped50ford" Taylor
"Plankowner", USS Asheville, SSN-758

Happily married, with seven wonderful children, we live in Jerusalem. I'm a software developer by profession. My PC gaming experience began back in 1998 when I downloaded Novalogic's demo for their first Delta Force game title. Later I became an addict of Operation Flashpoint. I hosted and was the main author of The Avon Lady's OFP FAQ. I came across Subsim after I had purchased my first naval sim, Enigma Rising Tide. I really enjoyed ERT and it led me to buy SH3 when it was released in early 2005. That's when I joined Subsim, seeking out the help of fellow kaleuns to increase my tonnage scores and evade those DDs. I played SH3 extensively for a bit less than two years and was an active member of Wolves at War. I have SH4 now but life's hectic and, as it is, I use too much of my spare time to blab on Subsim.

"Avon Lady"

I trolled Subsim for years before first posting sometime in 2005. It has played kind of a revolutionary role for me; prior to really getting involved with contributing, I had an amateur interest in maritime affairs and the navies of the world. Now I'm trying to make a career out of it! While I now spend most of the time writing enormous essays that only the ardent sub simmer might appreciate, memories of late-night Dangerous Waters multiplayer sessions are deeply etched in my mind: the hours of boredom trawling the oceans for any sign of the enemy (and occasionally falling asleep in the process) punctuated with moments of sheer terror when some faint low-hertz lines on the narrowband reveal themselves to be an inbound torpedo! Thanks, Subsim!

J. Matthew "fatty" Gillis

In the early '60s I had a game called Sub Battle. A molded plastic box that looked like a radar screen. Each player would set up their side (much like the game Battleship) with small plastic subs, and a few mines as well. The game had a moveable control arm, under the translucent radar screen, that moved a pointer. When a player decided to make a depth charge attack, he would push the center of the control arm down. If a sub was there, a light would indicate a hit. The light would also tell you if you struck a mine. The best thing about the game was the small periscopes on the corners. You could look under the top screen to see how close your opponent was coming to a mine, or maybe your last sub. I remember a lot of trash talking during gameplay with those periscopes. Next was Avalon Hill and the games Midway, Jutland, Wooden Ships & Iron Men. First computer game was Epyx's Sub Battle Simulation. I love this stuff!

Lee "CapnScurvy" Crawford

I found subsim.com by looking on the official Silent Hunter 3 website for communities and couldn't miss that big Subsim.com logo. I'm a member of this fantastic forum. I also came in touch with the wonderful mods for Silent Hunter 3, which made me enjoy the game even more. Also with the new Flightsim section added to the forum I think Subsim.com became the place I will be hanging around for a long time, as the website now contains the key genre's of simulations out there. For any simfreak out there, this is the place to be.

Wim "HunterICX" van Gestel

Ever since my youth, diesel subs have been my home. The very first Silent Hunter was my first taste of something that would grow to near obsession. Twelve years later at the age of 25, I've become an old salt and I found Subsim when I needed to find comrades as well as improving the unfortunate Silent Hunter II. Years passed but at last my calling with Silent Hunter III and IV have arrived, But I never ventured past World War II submarines because I'm old school. I need the smell of diesel. My comrades at Subsim are my neighbors as well as my brothers. We all have salt in our veins, diesel in our blood, and torpedoes as toys. Clear Horizons and good hunting.

Francis 'U-96' Bisaillon

My first ever Sub simulator was Silent Service on my brother's Commodore 64. I was 5 years old. I didn't really know what I was doing, but remember scoring few kills, possibly friendlies. Then I had a break from subsims, only to return when SH2 was released. At the time, I didn't have Internet, so I had to learn everything myself, which... sucked, so I quit SH2. My first visit to Subsim.com had to be somewhere around the release of SHIII. I was seeking for any info about the game and stumbled upon this marvelous site. A long time lurker, I decided to register on April of 2005 and felt like home from the moment I pressed the login button.

Teemu "Dowly" Siekkinen

I've had an interest in submarines for some time, I think it stems from a book I had as a child explaining all about submarines, how they work. I recall being fascinated by the shape of the Typhoon SSBN. About the same time I borrowed Silent Service II for the original Nintendo (NES) system, and spent many a happy evening torpedoing Japanese merchants before being told to go to bed. I bet O'Kane never had to fight Primary School as well as the Japanese….

After a few years shore leave, I returned to the silent service with Fast Attack in the late 1990s, commanding a 688 boat against the hordes of random tankers and oil rigs that got in my way. Occasionally I'd manage to find and hit an enemy boat, although usually the only submarine I sank was my own. However, by the turn of the century I'd gotten older and wiser and when I purchased Sub Command and Silent Hunter II, I was finally able to take my boat out into enemy waters and persecute the enemy!

The Subsim community is truly a great place to be. We're all united in our love for all things submarine, and it's that common bond, carefully nurtured by 'Die Onkel' that brings us together, be it in campaigning for a new patch, or mourning the loss of a fellow member. United, we are Subsim.

Jamie "Oberon" Currie

I love submarines. I have always loved submarines, even when I was 11 years old. I use to watch the old black and white series called, *Silent Service*. Served nine years US Navy with six years on diesel and nuclear submarines. After leaving the US Navy I never again had relationships like I did on submarines. Just something about men living a life of danger, a life where you have to trust the men on watch when you're off watch, a life of going without many things others take for granted, a life of excesses after 90 days at sea with no sun light.

How do you replace a lifestyle of submarine service? I was playing a subsim back in 1999 called Silent Hunter, an old WWII US Navy subsim set in the Pacific during the years 41-45. I did a word search on a search engine, I don't even think they had Google in those days, anyway I found Subsim.com.

I expect to introduce my grand children to Subsim next year … then I can have someone to razz me that I can get even with!

Ray (geetrue) Armstrong

ret STS 2 (SS)

Contributors

Zeb Alford was born in McComb, Mississippi, in 1925. He entered the Naval Academy in 1944. After graduation from Submarine School in 1949, he served aboard three-diesel submarines and three nuclear submarines. Zeb joined the nuclear Navy in 1958 when selected by Admiral Rickover, the father of the nuclear Navy. After nuclear school, his first nuclear assignment, as Executive Officer, was building and commissioning the first nuclear submarine designed to find and kill other nuclear submarines, *USS Tullibee*. He attended Charm School (Admiral Rickover's staff) for six months; then, as a Lt. Commander, he reported in 1962 as Commanding Officer of the nuclear attack submarine *USS Shark*. *Shark*, the fastest submarine ever built at that time. As a Navy Commander in 1965, he became the Commanding Officer of the Polaris nuclear submarine USS Sam Houston. He became Executive Assistant to the Undersecretary of the Navy (now Senator John Warner, (R-Va.). Zeb attended the National War College in 1970-1971 and reported as Commanding Officer, Guantanamo Bay Naval Station, Cuba, in 1971. Zeb retired from the Navy in 1973.

Zeb's second career has been in energy. Since entering this field he has been president of three different energy companies. In 1984 Zeb married Joan Chasan of Houston, Texas, and Long Island, New York. He has four grown daughters from a previous marriage. In 1985 he started a new company, Dolphin Energy, Inc., and began to market natural gas to major utilities on both coasts. Zeb is currently active in tennis, the Navy League, the Naval Academy Alumni Association, the Naval Order, and the Republican Party. In 1999, Zeb became Chairman of the *Cavalla* Historical Foundation which raised money to restore the WWII submarine *USS Cavalla* (SSK 244) and the DE-238 *USS Stewart*, both located at *Seawolf* Park, Galveston Bay, Texas. Thanks in large part to his leadership, they both have been restored and are open to the public for visits.

Robert Dexter Armstrong was born in Rome, Georgia on the day after Christmas, 1940. He qualified in submarines in 1960 and served six years served aboard *USS Drum* (SS-228), *USS Diablo* (SS-479), and *USS*

Requin (SS-481). Attended University of South Carolina, American University, Washington, DC and University of Oslo, Norway. Worked as statistical draftsman at the International Monetary Fund, Washington, DC, senior field representative for the Potomac Electric Power Company, sales engineer for the Haughton Elevator Company. Mr. Armstrong retired at thirty years as the Deputy Director of Space Acquisition and Building Management of the U.S. General Services Administration (GSA). Wife, Solveig Elise Nordvik (Armstrong) died December 13, 2005. Presently retired and loafing and spends time with daughters Kristine Margaret and Catherine and three grandchildren; Andrew, Grace-Solveig, and Calvin a.k.a. "Bulldozer".

Alan "Chock" Bradbury started out as a graphic designer and then switched to being a writer, before deciding that working freelance doing both these things would be more fun. He also trains people in these fields. With an interest in aircraft as well as submarines, he is a qualified pilot, but spends most of his time playing with his dogs and arguing with his wife, Maxine. Usually about whose turn it is to cook dinner. They live in Hazel Grove in the north of England.

Spencer Burnham is a freelance graphic designer/3D artist who lives with his wife Lisa and son Calvin in Barrington, New Hampshire.

Craig "Torplexed" Dinkelman was born May 12, 1962 on the old USAF Itazuki AFB in Fukuoka Japan. Lived on the Karamursel AFB Turkey, Kadena AFB Okinawa, and Fort Meade, Maryland as a kid. Despite being "an incorrigible Air Force brat", he developed a strange affinity for the naval stuff. Mr. Dinkelman attended the Praxis Art School and the Art Institute of Seattle and learned how to draw on paper with ink and pencils. He moved on to running the art department of a small Tacoma firm, now drawing with computers.

Jim Frantz is a former U.S Navy submarine Commanding Officer. He served on *USS Stonewall Jackson*, *USS Haddock*, *USS Flasher*, *USS Pattrick Henry* and *USS Barb*. Since leaving the Navy in 1985, he has worked in the software industry. He was a member of the Norton Utilities team before moving to Software Sorcery to try his hand at computer games. Mr. Frantz is currently a senior software engineer at Northrop Grumman Corporation.

J. Matthew Gillis is an undergraduate at King's College in Nova Scotia, Canada. His interests lie largely in defense policy, especially in topics of Canadian maritime security. Matthew has received two Bruce Oland awards for essays on Canadian Arctic sovereignty and naval humanitarian missions, and a Naval Memorial Trust prize for a paper on Allied corvettes in the Second World War. He resides near Halifax.

Andrew Glenn grew up in Switzerland and Australia where he worked for the Australian Department of Defence. In his capacity as a senior policy adviser on international defense issues, he helped establish a professional defense force for the newly independent nation of East Timor. Later he worked as a senior intelligence analyst on strategic issues. For a change of pace Andrew moved to Vancouver, Canada, where he is now working on his first novel.

Ron Gorence served on *Razorback*, *Bashaw*, *Swordfish*, *Tang* and *Sabalo* in the '50s and '60s, and was a Recruit Company Commander at RTC San Diego '62-65,'72-75. Retired as QMC(SS) and earned a BS at SDSU and then worked at General Dynamics, Electronics Div. for fifteen years and retired again. Married to Mary Ann and San Diego for 43 years and counting.

Tammy Goss is a PhD candidate in Linguistics at University of Wisconsin in Madison where she specializes in studying Native American Endangered Languages. Tammy became interested in submariner speech from hearing her husband Ken use Navy jargon at home. Kenneth served aboard *USS Louisville* SSN 724 from 1992 to 1996. Tammy and Ken live in Madison and in addition to working and studying, they enjoy spending time with their grand-daughter, Sahmiyah.

Sir William Stephen King-Hall, Baron King-Hall, usually went by his middle name of Stephen. He was born on 1 January 1893, the son of Admiral Sir George Fowler King-Hall and Olga Felicia Ker. He married Kathleen Amelia Spencer, daughter of Francis Spencer, on 15 April 1919. He was educated at Lausanne, Switzerland and at Royal Naval College, Dartmouth, Devon, England. He fought in the First World War between 1914 and 1918, with the Grand Fleet and 11th Submarine Flotilla. He gained the rank of Commander in the service of the Royal Navy. He held the office of Member of Parliament (M.P.) for Ormskirk Division between 1939 and 1945. He was invested as a Knight Bachelor in 1954. He was created Baron King-Hall, of Headley in the County of Hampshire on 15 January 1966. In addition to his anonymously written novella, *Diary of a U-boat Commander*, Sir Stephen authored *Submarines in the Future of Naval Warfare* in 1920. He died on 2 June 1966 at age 73.

Mike Hemming qualified on *USS Requin* SS-481. Then, after flunking out of Nuclear Power School, he returned to Subron 6 to spend four years on *USS Carp* SS-338. He left the service in '67 as a MM1 (SS) to return to his home town after finishing his BS at the University of Maryland. He has worked since at his family's nursery on the Eastern Shore of MD. In 1990 he became interested in finding old shipmates and started the *USS Carp's* reunions. At the same time he started to write about the submarine service. He lives with his wife of 42 years, Flo. They have one son and two grandchildren.

Don Meadows knew even at an early age he had to be a submariner. After high school he enlisted in the US Navy. He served aboard *USS Ray* SSN 653, *USS Miami* SSN 755, *USS Springfield* SSN 761, and *USS Dallas* SSN 700. While on board he began his first novel, working on the manuscript while the ship was submerged. The result was *Of Ice and Steel*, a well-received novel about a German submarine making a dangerous passage into the Arctic ice with an experiment that would guarantee survival of the Third Reich. He lives in Charleston South Carolina.

Kevin Moffat has been married to Donna for 17 years and have three kids. They live in Scotland, about 25 minutes from the old *USS Hunley* Naval Base in the Clyde. He has been drawing seriously now for about six years and specializes in aviation art. He found it was a real nice change to draw the tower of a Gato. He played the Silent Hunter series since SH1 and followed each version until the next came out.

Don Murphy served in the Admin Department and as Ship's Photographer onboard one of the original "41 For Freedom" nuclear missile submarines, the *USS Casimir Pulaski* (SSBN 633). Don served during the Cold War in the mid 1980's and retired in 1989. He is married to Lisa Dibley and they live in Florida with their three children, three dogs and one bird. Don is the webmaster of www.usscasimirpulaski.com and in his spare time builds submarine models.

Mariano "Marcantilan" Sciaroni (31), graduated as a lawyer in 1999 and specialized in corporate and commercial law. He has also spent two years in the Higher War School, Argentine Army, attending the Master's Degree Program on Strategy and Geopolitics. Mr. Sciaroni lives with his wife in Buenos Aires, Argentina.

Neal Stevens enjoyed reading about submarines and U-boats as a boy. His favorite pre-computer games were Stratego, chess, Battleship, and Risk. He created his own submarine wargame, which consisted of a two-decker cardboard playing surface with numbered grid, a deck of cards, five dice, and six sheets of outcomes. He started a website in 1997 with six submarine game reviews and called it Subsim.

Valerie Stevens is an architect major at Texas A&M University. At age nine, she was among the first volunteers to restore the *USS Cavalla* in Galveston, TX. She excels at tennis, art, and the lost cause of rooting for the Oakland Raiders.

Shawn Storc accidentally found his way into the computer/video game industry after spending way too much time in college having fun and avoiding the 'real' world. Amazed that he still has a job after all these years in the business, he has worked on many diverse, memorable and sometimes not so memorable products. In his free time he raises hell with his lovely, patient wife Nicole and their menagerie of rescued pets and motorcycles.

Grant Swinbourne has been an IT professional for over 20 years. Currently working as Manager, IT Strategy for Qantas Airways, he has held a number of key IT roles in the 10 years of working for the airline. Most notably as Manager, Online Channels for Jetstar Airways when that airline started up. With a father who served in the RAN, he has always been keenly interested in all thing naval and in particular in naval aviation and submarine operations. He lives with his wife and 3 children in Sydney Australia.

Chris "Lobsterboy" Weisensel was married this year to his wife, Theresa. During scuba lessons for their honeymoon, he discovered he had an unfortunate allergy to neoprene. Once in Hawaii she found casual amusement in his obsession with all things naval. He has a B.A. in international studies and a minor in psychology. They reside in a suburb of Minneapolis.

Katana

Acknowledgments

Special thanks for assistance in numerous ways with this edition of the Almanac to Bernie Tyler (Executive Producer of Fast Attack), Joe Buff, Bill Parker, Judy Heise, and Dr. Erica J. Benson of the University of Wisconsin-Eau Claire, George Strake, Herbert Taylor and the Galveston County Daily News.

Art contributors include Torplexed, Bobo, CCIP, Katana, Sulikate, Orret, and many others.

A salute goes out to the Silent Hunter 4 people of Ubisoft: Dan Dimitrescu, Cristian Hriscu-Badea, Chris Easton, Francisco De la Guardia, Ioan Manea, Elvin Gee, Peter Dalope, Tudor Serban, Olivia Zaharia, and the rest of the Romanian dev team. Keep up the good work.

A hearty job well done as only the U.S. Navy can do it to Captain John Litherland, Petty Officer Michael "Ping Jockey" Granito, Cdr. Jim Gray, Master Chief Larry Batten, Cdr. Al Onley, Senior Chief Richard Lattimer, Senior Chief Marty Ledesma, PAO Phil McGuinn and Chris Lounderman, and the crew of the *USS Texas* SSN-775. *Don't mess with Texas!*

Thanks also go to Mark Hooper at Angel Editing, Rachel Owens, and Theresa Flanders for assistance in proofing the texts.

With much sadness we note the passing of Abraham Zeegers, a friend and colleague. Abraham was a well-respected forum moderator who brought a good deal of wisdom to our discussions. We will always remember his well-prepared tour of the delta works during the 2006 Subsim Meet in Holland. We visited the Maritime museum in Amsterdam and the replicas of the *Batavia* and the *Amsterdam*. Abraham was loyal, considerate, and a genuine pleasure to know. Farewell, friend.

Neal Stevens

I was a regular back when the oceans were gray and the weekend wars between Red and Blue opponents of Iron Wolves kept a person enthralled in front of a monitor. I found myself led eventually to the forums of Subsim. In search of more and better adventures I have sailed the oceans of Enigma and SH 3&4. Experimented with the seas of BattleStation Midway and Navy Field. Withstood the deadly storms of Pirates of the Caribbean and Voyage Century. The forums of Subsim have been a port of refuge in between the many different moods of the seas of naval gaming and I thank Neal Stevens and all the other captains of the world for being part of such an elite and unique community of people.

Robin "Iceman" Hawkins

SURFACING NOVEMBER 2008

2009 SUBMARINE ALMANAC

When surfing the web, sail over to Subsim.com

Web's #1 submarine and naval game resource.

www.subsim.com

www.ingramcontent.com/pod-product-compliance
Lightning Source LLC
Chambersburg PA
CBHW032039090426
42744CB00004B/60